YO-BQH-741

Power Supplies for Electronic Equipment

Volume 1

Power Supplies for Electronic Equipment

Volume 1 Rectifiers, Inverters and Converters

J. R. NOWICKI

C. Eng., F.I.E.R.E., S.M.I.E.E.E.

A DIVISION OF
THE **CHEMICAL RUBBER** CO.
CLEVELAND, OHIO

INTERNATIONAL SCIENTIFIC SERIES

Published by
Leonard Hill Books
a division of
International Texbook Company Limited
158 Buckingham Palace Road, London SW 1

First published 1971

Library of Congress Catalog Card Number: 73-148082

This edition published 1971 by
The Chemical Rubber Co.
18901 Cranwood Parkway
Cleveland, Ohio 44128,　USA

Printed in Great Britain by
J. W. Arrowsmith Ltd., Bristol and London.

To my wife Zonia for her help,
encouragement, and forbearance.

PREFACE

The power supply is an essential part of every electronic equipment. In its simplest form it may consist of no more than a transformer, rectifier, and smoothing circuit, but frequently much more sophisticated arrangements are required, especially in the industrial field of computers, digital instruments, d.c. amplifiers, etc.

Since the introduction of transistors and other semiconductor devices in in the late 1940s, the interest in all types of power supplies, including d.c. inverters and converters, has grown considerably. The advantages of ruggedness and higher overall efficiency in using semiconductor devices when compared with the earlier valve counterparts are well known, and are particularly beneficial as greater emphasis is now placed on reliability, size and weight reduction, and portability.

Numerous papers have been published, often dealing with one particular aspect of the subject. Many of the references are not readily available and tracking down information often proves a time-consuming undertaking. While working in the field during the past twelve years, I have frequently been faced with the unenviable task of wading through vast amounts of material in order to extract the required reference.

These two volumes, therefore, are an attempt to present up-to-date available material and to give the necessary references. The basic theory is supported by circuit analysis and, in many cases, is followed by a detailed design procedure. Many practical examples are given to provide the reader with reliable and ready-to-use circuits.

They aim to supply the need for a comprehensive study of the subject for the use of all grades of electronic engineers, technicians, and students at universities and technical colleges.

Patent Protection

Some of the circuits, semiconductor devices, and arrangements described here are subject to Patent protection. Anybody wishing to make use of the above should obtain the permission of the Patentee.

JR Nowicki

ACKNOWLEDGEMENTS

I wish to thank the Directors of Mullard Ltd. for making the publication of this book possible by granting permission to use much of the material from the Company's publications.

I am grateful to all the following for permission to use their material: Institute of Electrical and Electronic Engineers, Institution of Electronic and Radio Engineers, Institute of Physics and The Physical Society, Instrument Society of America, A.T.E. Journal, Bell Laboratories Record, Bendix Corporation, Control Engineering, Delco Radio, Design Electronics, Direct Current, EDN Electronics Design News, EEE Circuit Design Engineering, Electronic Applications, Electronic Components, Electronic Design, Electronic Engineering, Electronic Equipment News, Electrical Manufacturing, Electronic Products, Electronics, Electro Technology, Electronics World, Elektronik, Elektronische Rundschau, Ferranti Ltd., General Electric Company (and International General Electric Co. of New York Ltd.), Hewlett Packard, Industrial Electronics, Instrument Practice, International Rectifier (and International Rectifier Company (Great Britain) Ltd.), Kepco, Light and Lighting, McGraw-Hill Book Company, Miniwatt, Minneapolis - Honeywell Regulator Company, Motorola Semiconductor Products, Naval Research Laboratory (USA), Philips, Physical Review, Physics Review, Pitman, Proceedings of Royal Society, Radio Mentor, Royal Aircraft Establishment, Radio Corporation of America (and RCA Limited), Review of Scientific Instruments, Semiconductor Products, SGS Fairchild, Silicon Transistor Corporation, Solid State Design, Telefunken, Texas Instruments Inc., Westinghouse Electric Corporation, Westinghouse Brake & Signal Co. Ltd., Wireless Engineer, Wireless World, Zeitschrift fur angewandte Physik, and any person, publication, or organisation that has in any way contributed to this book.

Finally, I would like to express my gratitude to all those who have helped with the preparation of this book, and in particular to Mr. D. F. Grollet for supplying material on 'transistor switching characteristics' included in Chapter 1, and Mr. M. J. Endacott for reading the manuscript and offering constructive criticism.

CONTENTS

SYMBOLS

a or A	anode terminal
A	cross-sectional area of core
av or AV	average
b or B	base terminal
B	flux density
B_M	maximum operating flux density
B_s	saturation flux density
BO	breakover
BR	breakdown
c or C	collector terminal
c or C	capacitance
$C_{b'c}$	transistor base–collector capacitance
$C_{b'e}$	transistor base–emitter capacitance
C_o or C_{out}	output capacitance
C_{TC}	capacitance of collector depletion layer
C_{TE}	capacitance of emitter depletion layer
CC	constant current
CV	constant voltage
d	delay or duty cycle
D	diode
e or E	emitter terminal
e	instantaneous voltage
E	applied voltage
E_{dc}	d.c. output voltage
E_{max}	maximum applied voltage
E_s	energy stored
E_t	transferred energy
E_D	forward voltage drop across thyristor
E_K	voltage drop due to copper loss
E_T	transformer output voltage
$E_{T(max)}$	maximum sine-wave output voltage of the transformer
$E_{T(rms)}$	r.m.s. value of the transformer output voltage
f	frequency
f_{low}	low frequency
f_{max}	maximum frequency of oscillations

f_o	optimum frequency
f_r	ripple frequency
f_T	transition frequency (common product of emitter gain and bandwidth)
f_1	frequency of unity current-transfer ratio modulus
g *or* G	gate terminal
g_m	mutual conductance of transistor
G_m	mutual conductance of stage
h_{FB} and h_{FE}	static value of forward current-transfer ratio with output held constant
H	henry
H	magnetising field strength
H_s	value of magnetising field strength at saturation
H_0	intrinsic strength of magnetising field
Hz	hertz
i	instantaneous current
i_{av}	average value of a.c. current
i_{pk}	peak value of a.c. current
i_r	instantaneous reverse current
i_{rms}	r.m.s. value of a.c. current
i_C	instantaneous value of capacitor current
i_F	instantaneous forward current
I_{av}	total average current
$I_{b(min)}$	minimum base current
$I_{b(pk)}$	peak base current
I_c	r.m.s. value of collector current *or* total capacitor current
$I_{c(rms)}$	r.m.s. value of capacitor current
I_{dc}	d.c. value of total current
I_i	input current *or* inverse current
$I_{i(max)}$	maximum inverse current
I_m	magnetising current
I_{mag}	r.m.s. value of transformer primary magnetising current
I_o *or* I_{out}	output current
I_{on}	initial switch-on current
$I_{o/c}$	sum of magnetising current and core loss components of transformer with either primary or secondary open-circuited
I_{pk}	peak current
I_{rms}	r.m.s. value of current
I_B	base current
$I_{B(on)}$	base current of saturated transistor
$I_{B(off)}$	reverse base current during switch-off transition
I_C	total collector current
I_{CBO}	collector cut-off current (emitter open-circuited)

I_{CEO}	collector cut-off current (base open-circuited)
I_D	diode current
I_E	emitter current
I_f	feedback current
I_F *or* $I_{F(AV)}$	forward current *or* average forward current
I_{FG}	thyristor forward gate current
I_{FGM}	thyristor peak forward gate current
I_G	thyristor gate current
I_H	thyristor holding current (d.c.)
I_{IN}	average supply current
I_L	load current
I_L	thyristor latching current
I_L	value of inductive current
I_M	magnetising current
I_R	continuous d.c. reverse leakage current
I_R	current flowing through the resistor R
	or collector load current
I_T	thyristor continuous (d.c.) on-state current
$I_{T(AV)}$	average value of anode current
I_V	tunnel diode valley point current
I_Z	current through voltage regulator diode after breakdown
I_{Zs}	specified current through voltage regulator diode after breakdown
K	constant
l	length of magnetic path
l_c	length of flux path in core
l_g	length of air gap
L	inductance
L_{crit}	value of critical inductance
L_p	inductance of primary
L_t	inductance of winding N_t
n	number $1, 2, 3, ..., n$
N_b	number of turns in base winding
N_f	number of turns in feedback winding
N_h	number of turns in heater winding
N_i	number of turns in ignition winding
N_p	number of turns in primary winding
N_s	number of turns in secondary winding
N_t	number of turns in control winding
p	percentage change
P	steady-state dissipation
P_c	collector dissipation
$P_{c(max)}$	maximum collector dissipation

$P_{c(transient)}$ collector transient dissipation
P_f power delivered by feedback winding
P_i input power
$P_{i(av)}$ average input power
P_K transformer copper loss
P_o or P_{out} output power
P_p pulse power
$P_{p(max)}$ maximum permissible pulse power
P_s steady-state dissipation
$P_{s(max)}$ maximum permissible steady-state dissipation
$P_{tot(max)}$ maximum total dissipation
P_F forward power loss or total power absorbed by drive circuit
$P_{F(AV)}$ average forward power loss
P_R power dissipated in resistor R or power drawn from the supply by bias chain
Q charge or charge remaining in the device after time t
Q_e extracted charge
Q_f charge extracted during forward recovery time
Q_i initial charge
Q_m or Q_{max} maximum charge
Q_{min} minimum charge
Q_r charge extracted during reverse recovery time
Q_t total charge extracted
r_b transistor base resistance of equivalent T circuit
$r_{bb'}$ internal base resistance of transistor
r_e internal emitter resistance of transistor
r_p or R_p winding resistance of transformer primary
r_s or R_s winding resistance of transformer secondary
r_{tot} total winding resistance of transformer
r_B base resistance of unijunction transistor
r_{BB} interbase resistance of unijunction transistor
r_Z dynamic resistance of voltage regulator diode
r_{Zs} dynamic resistance at specified current
R resistance
R_b or R_B external base resistance
R_{bb} sum of internal and external base resistances
R_{ext} external circuit resistance
R_o transistor input resistance obtained by drawing tangent to input characteristics or output resistance
R_{th} thermal resistance
$R_{th(c-a)}$ thermal resistance case-to-ambient
$R_{th(effective)}$ effective thermal resistance
$R_{th(h)}$ thermal resistance of heat sink

$R_{th(i)}$	contact thermal resistance
$R_{th(j-a)}$ or $R_{th(j-amb)}$	thermal resistance junction-to-ambient
$R_{th(j-c)}$ or $R_{th(j-case)}$	thermal resistance junction-to-case
$R_{th(j-mb)}$	thermal resistance junction-to-mounting base
$R_{th(s)}$	steady-state thermal resistance
$R_{th(s-r)}$	thermal resistance for permissible temperature rise
$R_{th(t)}$	transient thermal resistance
R_B	equivalent transistor input resistance
R_{BX}	total input resistance of compound transistor
R_{CE}	collector-emitter resistance of transistor
$R_{CE(sat)}$	saturation resistance of transistor
R_G	thyristor gate resistance
R_L	load resistance
R_V	variable resistance
S	stabilisation factor
S_p	stabilisation factor of pre-stabilising stage
S_F	fractional change coefficient
S_T	total temperature coefficient
S_{TR}	temperature coefficient of transistor
S_Z	temperature coefficient of voltage regulator diode
SCR	thyristor
SW	switch
t	time
t_d	delay time
t_{com}	commutation period
t_{cond}	conduction period
t_f	fall time
t_{fr}	forward recovery time
t_{off}	turn-off time or duration of off time
t_{on}	turn-on time or duration of on time
t_p	pulse duration or time of half-cycle
t_r	rise time
t_{rr}	reverse recovery time
t_s	storage time
T	transformer
T	temperature or periodic time
T_a or T_{amb}	ambient temperature
$T_{amb(max)}$	maximum ambient temperature
T_c or T_{case}	case temperature
T_{eq}	equivalent time
T_j	junction temperature

$T_{j(max)}$	maximum junction temperature
T_{mb}	mounting base temperature
T_n	temperature of n degrees Kelvin
T_r	reference temperature
T_s	source temperature
T_{s-r}	permissible temperature rise
TR	transistor
U_p	utility factor of transformer primary
U_s	utility factor of transformer secondary
v	instantaneous value of voltage
v_{pk}	peak value of instantaneous voltage
v_F	instantaneous value of forward voltage
v_R	instantaneous value of reverse voltage
V_{bb}	voltage applied to base of transistor
V_{be}	minimum value of base-emitter voltage
V_{cc}	supply voltage
V_d	forward voltage drop across rectifier diode
V_f	feedback voltage
V_{fr}	forward recovery voltage
V_i or V_{in}	input voltage
V_i	ignition voltage
V_h	heater voltage
V_o or V_{out}	output voltage
$V_{o/c}$	open-circuit voltage
V_p	peak point voltage or primary voltage
V_s	secondary voltage
$V_{s/c}$	short-circuit test voltage
V_x	voltage across ballast reactance
V_{BB}	unijunction interbase voltage or d.c. base-supply voltage
V_{BE}	base-emitter voltage
V_{BEM}	maximum base-emitter voltage
V_{BO}	breakover voltage
$V_{(BR)}$	breakdown voltage
$V_{(BR)CBO}$	breakdown voltage collector-to-base (emitter open-circuited)
$V_{(BR)CEO}$	breakdown voltage collector-to-base (emitter and base short-circuited)
$V_{(BR)R}$	reverse breakdown voltage
V_C	collector voltage
V_{CE}	collector-to-emitter voltage (d.c.)
$V_{CE(pk)}$	peak value of collector-to-emitter voltage
$V_{CE(sat)}$	collector-to-emitter saturation voltage

V_{CEM}	maximum rated peak collector voltage
V_D	forward voltage drop of p-n junction *or*
	forward voltage drop of rectifier diode
V_E	emitter voltage
V_{EB}	emitter-base voltage (d.c.)
V_F	d.c. forward voltage
V_L	voltage across lamp
V_R	d.c. reverse voltage *or* ripple voltage
V_{RR}	applied repetitive peak reverse voltage
V_{RRM}	repetitive peak reverse voltage
V_{R_p}	voltage drop across resistance of primary winding
V_{RB}	voltage drop across external base resistor
V_{RS}	voltage drop across resistance of secondary winding
V_{RSM}	maximum non-repetitive reverse voltage rating
V_{RW}	crest working voltage rating of rectifier diode
V_{RWM}	crest (peak) working reverse voltage
V_T	thyristor voltage between anode and cathode
V_Z	voltage across voltage regulator diode after breakdown *or*
	voltage regulator (Zener) diode operating voltage
V_{Zs}	specified reference voltage at specified current I_{Zs}
V_0	intercept voltage of tangent to forward characteristic
VA_s	secondary volt-ampere rating
W	watt
$W_{o/c}$	transformer copper loss and core loss open-circuit test
$W_{s/c}$	transformer copper loss and core loss short-circuit test
W_R	reverse switching transient power loss
X_L	reactance of ballast choke
α	turns for 1 mH (Ferroxcube cores)
β	h_{FE}, transistor current gain
δ	differential
η	efficiency
η_f	efficiency as function of frequency f
θ	angle in degrees
μ	permeability of core material
τ	time constant *or* rise time
τ_s	carrier storage time coefficient of switching transistor
ϕ	magnetic flux *or* angle in degrees
ϕ_{pk}	peak value of magnetic flux
ϕ_s	magnetic flux at saturation
ω_s	angular frequency, $2\pi f$
ω_t	product of gain and bandwidth
Ω	ohm

CHAPTER 1. SEMICONDUCTOR DEVICES

Silicon rectifier diodes, voltage regulator (Zener) diodes, and transistors are the essential components in modern electronic equipment.

The purpose of this chapter is to provide the reader with sufficient information on semiconductor devices used in the design of power supplies. It describes the principles of the operation and the main characteristics of these devices. Other devices which may be used are briefly described at the end of the chapter.

Silicon Rectifier Diodes

Characteristics

Silicon diodes with their low forward voltage drop and high reverse voltage characteristics provide almost ideal means of rectification.

The forward and reverse current–voltage characteristics of a typical silicon diode are shown in Figure 1 (Refs. 1 to 10). From the graphs it can be seen that, below a certain forward voltage V_{F_1}, only a very small current flows through the diode. Initially the current increases exponentially to the first approximation. When the voltage V_{F_1} is reached, ranging between 0·5 to 1·5 V, a forward breakdown occurs and the characteristic becomes almost linear. The current that flows in this range is limited by the differential resistance of the diode. Owing to the voltage drop across the differential resistance of the diode there is a certain amount of heat dissipation in the forward direction which cannot be neglected. In some cases provision for cooling may be necessary but this will be discussed later.

The reverse characteristic of the diode shows that the diode presents a very high impedance, so that only a very small current can flow in this direction. The current increases slightly with voltage until a breakdown

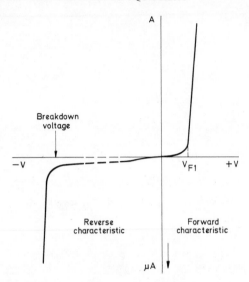

FIGURE 1.   Forward and reverse characteristics of silicon rectifier diode.

occurs.   If this stage is reached, any further increase in the reverse voltage
will cause a large reverse current to flow, which may lead to a deterioration
of performance or even to complete destruction of the device due to a very
high power dissipation.

A closer look at the forward and reverse characteristics will now be
made with reference to a typical silicon rectifier diode BYX38/600R, which
is intended for use in power rectifier applications up to 2·5 A mean forward
current at $T_{mb} = 125$ °C (Ref. 1).   The forward characteristics of the
BYX38 silicon rectifier diode series are shown in Figure 2.   Usually two
curves are given: a typical curve indicating a general trend for most
devices of the type, and a maximum spread curve indicating the limiting
values for the device.   The maximum spread curve shows that at 2·5 A the
maximum forward voltage drop is 1·55 V giving a maximum value of resistance of approximately 0·292 Ω, while a typical value is about 0·216 Ω at
the junction temperature of 125 °C.

The reverse characteristics of the diode are shown in Figure 3.   The
curve for the maximum spread shows that at the reverse voltage of 400 V
the maximum forward current is 200 μA.   Therefore the minimum value
of the diode resistance in the reverse direction is 200 kΩ, whilst the maximum
diode resistance is 850 kΩ.

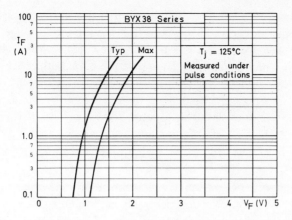

FIGURE 2. Forward characteristics of BYX38 diodes.

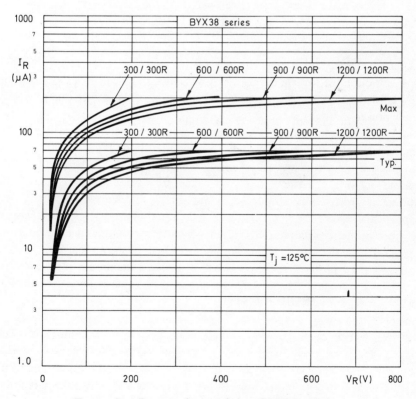

FIGURE 3. Reverse characteristics of BYX38 diodes.

If the characteristics of the silicon rectifier diodes are studied at various junction temperatures, it will be seen that the forward voltage drop across the rectifier decreases slightly with increasing temperature, giving rise to a lower ohmic component. In the reverse direction both the reverse current and the reverse voltage are temperature-sensitive and both increase with increase in temperature. Because of the change of the characteristics with temperature the curves shown in Figures 2 and 3 were plotted for a junction temperature of 125 °C. Using these curves in designing a rectifying circuit a worst possible condition is arrived at. Therefore in practice a much better performance can be expected.

There are a number of limitations regarding the reverse voltage ratings.

(i) The maximum non-repetitive reverse voltage rating V_{RSM} should never be exceeded by any supply transient. The maximum non-repetitive reverse voltage rating is less than the breakdown voltage of the diode.

(ii) The maximum repetitive peak reverse voltage rating V_{RRM} must never be exceeded by any repetitive voltages in the circuit, and their duration must not exceed the limit given in the data sheets.

(iii) There is a specified maximum crest working voltage V'_{RRM} which was previously known as the recurrent peak inverse voltage. This is the crest value of the same wave applied to the rectifier diode.

In addition to the above there are three forward current ratings for the rectifier diodes: the maximum mean forward current, the maximum repetitive peak forward current, and the maximum surge forward current. The actual figures for the above ratings are usually quoted by the manufacturers. The first two figures for the BYX38 diode are $I_{F(AV)}$ at $T_{mb} = 125$ °C of 2.5 A and I_{FRM} of 20 A respectively. The maximum forward one-cycle surge current is 38 A. Often additional information is given in the form of a graph plotted for the peak current against surge duration.

Power Losses in Rectifier Diodes

The three main sources of rectifier diode power loss (Ref. 11) are (i) forward voltage drop V_F when the rectifier diode is in its conducting state, giving rise to forward loss P_F, (ii) leakage current (also termed inverse current or reverse current) when the rectifier diode is in its blocking state, giving rise to inverse loss P_i, and (iii) switching transients when the rectifier diode is switched from one state to the other.

Forward Loss

When the rectifier diode is in its steady conducting state, there is a small voltage drop across the diode junction. The voltage drop increases with increasing current, but is independent of the frequencies considered in this section.

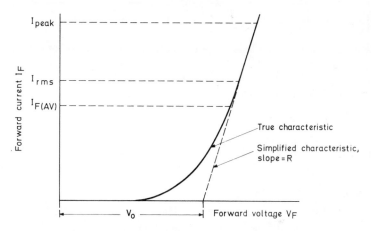

FIGURE 4.   Typical forward characteristic of rectifier diode.

If the normal characteristic of forward voltage drop against current (Figure 4) is considered and a tangent from the peak current I_{pk} back to the voltage axis, is drawn, calculations can be simplified by using this tangent instead of the actual curve. The forward loss P_F caused by this voltage drop can be calculated by means of the following equation (Ref. 11):

$$P_{F(AV)} = V_oI_F + I_{rms}^2R, \tag{1}$$

where I_F is the average forward current per rectifier diode, I_{rms} the r.m.s. current per rectifier diode over the whole cycle, V_o the voltage given by the intercept of the tangent on the voltage axis of the forward characteristic, and R the slope of the tangent to the diode forward characteristic.

The forward loss and the inverse loss together form the major part of the power loss at low frequencies.

Inverse Loss

When the rectifier diode is reverse-biased, it is in its high-resistance state, but a small leakage current still flows. The leakage current is sensitive to temperature, and it may double for every 10 to 15 degC rise in temperature.

The power loss due to leakage current is independent of frequency. The inverse loss at any temperature can be calculated from Equation (2) given below (Ref. 11). The dissipation loss calculated by means of this equation gives a slightly higher value than occurs in practice.

$$P_{i(av)} = V_{RWM} \times I_{i(max)} \times \frac{1 - \frac{1}{2}\phi \cos\frac{1}{2}\phi + (1/\pi)\sin\frac{1}{2}\phi}{1 + \cos\frac{1}{2}\phi}, \tag{2}$$

where V_{RWM} is the crest working voltage, $I_{i(max)}$ the maximum inverse current corresponding to V_{RWM}, and ϕ the conduction angle.

Switching Losses

The switching losses are caused by the forward switching transients and the reverse switching transients. (They are sometimes termed forward recovery transients and reverse recovery transients respectively.)

Forward Switching Transients. When a semiconductor rectifier diode is switched from the non-conducting state to the conducting state, it does not reach steady-state conditions in the forward direction instantaneously. The minority carriers (holes, in the case of p–n rectifiers) require a finite time to modulate the conductivity of the material in the vicinity of the junction, and thus to reduce the resistance of the device to the flow of current in the forward direction.   For a given current, the initial voltage developed across the rectifier is therefore higher than it is in the steady state (Ref. 12).

The forward switching transient is shown in Figure 5. The peak value of the forward recovery voltage V_{fr} depends on the rate of rise of the switching current.   The duration of this transient, which is termed the forward recovery time t_{fr}, is generally small compared with the reverse recovery time. The extra dissipation in the device is caused by the forward recovery voltage and the rising forward current.   The power loss due to the forward switching transient is normally much smaller than that due to the reverse switching transient, and it may be neglected for most types of rectifier diode.

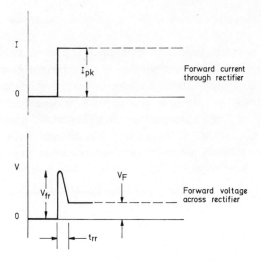

FIGURE 5.   Forward switching transient.

Reverse Switching Transients. For a steady-state forward current I_{pk}, there is a corresponding flow of minority carriers from the junction. The minority carriers require a finite time to recombine after the forward voltage is removed; therefore the capability of current flow does not cease instantaneously. The rectifier diode only attains its high-resistance state in the reverse direction when the minority carriers have recombined within the material.

The process of reverse recovery can be accelerated by applying a reverse bias to the rectifier diode. If the rate of rise of the switching voltage is high enough, the uncombined minority carriers will be extracted by the reverse voltage in the form of a reverse current spike. The total extracted charge is made up of two components. One component is due to the minority carriers, and the other to the diode capacitance, which gives rise to capacitive current in the process of charging.

The idealised form of the reverse recovery transient is shown in Figure 6. When the rectifier diode is switched from the forward to the reverse direction, a large transient current I_R can flow owing to the diffusion of the minority carriers back across the junction. This current is limited initially only by the applied reverse voltage V_R and the external circuit resistance R_{ext}. This condition will prevail, and the rectifier diode remain forward-biased, until the stored charge concentration at the junction boundary has fallen to zero, at point A. The rectifier diode resistance then builds up to its high value, and the reverse current decreases to the leakage current value under the normal blocking condition during the reverse recovery period t_{rr}.   Typical t_{rr} is 300 ns for a fast-recovery diode.

The major part of the power loss due to reverse recovery transients occurs while the rectifier diode resistance is building up to its high value. During this period the voltage across the junction is increasing to the applied reverse voltage V_R, and the peak reverse current I_R is decreasing to the normal leakage current.   The reverse transient loss increases with current and frequency.   The amount of dissipation caused by this transient is negligible at normal power frequencies, but it can be appreciable at higher frequencies.

Reverse Recovery Transient

The reverse recovery transients obtained in practical circuits differ from the idealised transient shown in Figure 6 (Ref. 13).   The reverse recovery time, the peak reverse current, and the charge extracted from the rectifier diode are dependent on a number of circuit parameters, as well as on the rectifier parameters.

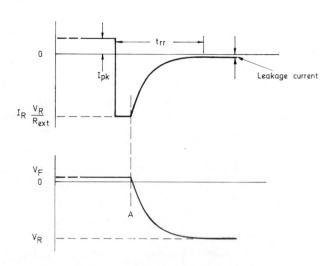

FIGURE 6.   Reverse switching transient.

For any particular type of rectifier diode, there is a spread in the rectifier diode parameters; therefore there is a spread in the charge extracted under specified conditions.

The rate of recombination of the minority carriers and the rate of application of the switching voltage are of special importance for estimation of stored charge under pulse conditions. If the switching time is comparable with the recombination time, then the extracted charge will be very different from the initial charge that existed in the device when the switching voltage was applied. The extracted charge will approach the initial charge in the device if the switching time is made much smaller than the recombination time.

The dissipation loss due to the reverse recovery transient will therefore vary with the type of circuit and the circuit conditions for any particular rectifier diode. It is, however, useful to know the dissipation due to reverse recovery transients under specified conditions. This information will enable the likely dissipation loss to be assessed in a particular circuit. The dissipation loss in the case of fast rectifier diodes like BYX30 and BYX46 is given in the published data (Ref. 9).

Voltage Regulator (Zener) Diodes

Voltage *regulator* diode is a more accurate name for the semiconductor device that has been known as a Zener diode; many of the devices do not, in fact, use the Zener effect. Voltage *reference* diode is a term adopted for voltage regulator diodes that have been designed for use in applications where the diode current is constant.

Voltage regulator diodes provide a source of stable reference voltages that has several advantages over the gas-filled stabiliser and reference tubes. The most important of these is the great range of reference voltages that the diodes make available; this extends from less than 4 V to more than 100 V, the particular voltage depending on the type of diode.

The voltage regulator diode is a silicon semiconductor device having a normal diode characteristic in the forward direction and a constant-voltage property in the reverse direction with a reverse current above a certain value. The precise value of this constant voltage is a property of the individual diode.

The range of voltage regulator diodes available is made up of several series, each having a characteristic power rating and offering a wide range of voltages. The diodes within a series have nominal voltages that follow the E24 logarithmic series of preferred values. The tolerance on these voltages is $\pm$ 5 $\%$ so that complete coverage within the total voltage range is offered.

Correctly exploited, this constant-voltage phenomenon makes voltage regulator diodes exceptionally useful devices for obtaining a specific voltage that is relatively undisturbed by fluctuations in the supply voltage or load current. Hence, the diodes find many uses in a great variety of voltage reference circuits (Ref. 14).

Voltage regulator diodes are silicon semiconductor devices that have voltage–current characteristics similar to those shown in Figure 7. For a given type of diode, the *knee* always occurs within close limits about the same voltage, the precise value being a constant for each individual diode.

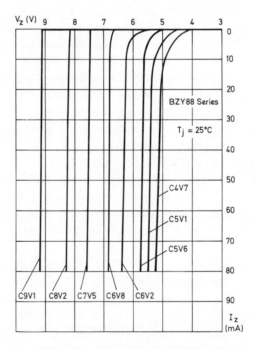

Figure 7. Typical dynamic characteristics of voltage regulator diode.

Zener Effect

When it is forward-biased, the voltage regulator diode behaves as a normal diode.   When the voltage across it is reversed, a leakage current of only a few microamperes flows through the diode.   This current is independent of the applied voltage over quite a wide range.   If the voltage is increased, it eventually reaches a value at which the current increases suddenly to a large value—tens of milliamperes or even several amperes. The current must be limited to keep dissipation within the power rating of the device.   The voltage at which this occurs is known as the *breakdown* voltage $V_{(BR)}$.

This phenomenon was called the *Zener* effect when first observed because it was thought to be due to the mechanism described by Zener (Ref. 15) in his theory of breakdown phenomena in dielectrics.   Today, this breakdown in voltage regulator diodes is thought to be due to field effects when it occurs at voltages below 5 V, and is attributed to an avalanche effect when it occurs at voltages above 7 V.   Between 5 and 7 V, breakdown is caused by a combination of the two effects.   Thus, Zener's breakdown theory remains valid for low voltages, and the avalanche theory proposed by McKay and McAffee (Ref. 16) is accepted for breakdown at high voltages.

When the reverse voltage across the diode is small, the current flowing is the surface leakage current plus the normal saturation current which consists of thermally generated holes and electrons.   At first, increasing the voltage does not increase the number of carriers and, therefore, cannot increase the current.   However, if the voltage is continually increased, it will eventually reach a value that can cause the spontaneous disruption of the covalent bonds among the atoms close to the junction.   (This is analagous to the phenomenon of cold emission from metallic surfaces in the presence of a high accelerating field.)   The disruption of the bonds results in the generation of hole–electron pairs that produce a great increase in the current. This action is the *Zener* effect.

Avalanche Effect

The voltage at which the Zener effect occurs depends on how the silicon is doped, and can be quite high.   Consequently, a sudden current increase is often achieved by means of an *avalanche* effect below the voltage level at which the Zener effect would otherwise occur.

If the covalent bonds are not broken, the current remains at only a few microamperes until the avalanche effect occurs.   As the voltage across the diode is increased, the velocity of the carriers is increased. Eventually, this velocity is sufficient to cause ionisation when electrons collide with the semiconductor molecules.

Breakdown Stability

Voltage breakdown in a regulator diode occurs at a definite, stable level which is practically constant, no matter which mechanism (Zener effect or avalanche effect) is the cause. Provided that the maximum permissible junction temperature is not exceeded, the breakdown is reversible and non-destructive. The voltage across the diode after breakdown is known as the *reference* voltage; this cannot greatly exceed the breakdown voltage and is virtually constant. The tolerances quoted in the published data (Ref. 9) indicate the voltage limits for a type of voltage regulator diode, not an individual device; each individual diode will have its own constant reference voltage within the limits quoted.

When breakdown occurs, the large current change is accompanied by a small voltage change. Therefore, the dynamic slope resistance r_z, changes from a high value to a very low value. Published data sheets usually quote the slope resistance at three different current levels.

Effects of Temperature Changes

Although the breakdown voltage is essentially a constant in voltage regulator diodes, voltage changes are caused by changes in current and junction temperature; the effects of temperature changes are minimised when testing the diodes by using pulsed voltages and currents. Consequently, the characteristics are obtained before the junctions get hot. Therefore, unless otherwise stated, the information in the published data applies when the junctions are initially at an ambient temperature of 25 °C.

The basic curve of the voltage regulator diode is its dynamic characteristic; this is the variation of reference voltage with pulsed current. The slope of this characteristic is the dynamic resistance r_z. Figure 8 compares dynamic characteristics at 25 °C and at another ambient temperature T_n. The displacement between the two curves is caused by the temperature sensitivity of the diode characteristics, and equals a voltage corresponding to $S_z(T_n - 25)$. where S_z is the temperature coefficient of the diode.

The temperature coefficient is negative in diodes with a breakdown voltage below 5 V, and positive in diodes with a higher breakdown voltage. Therefore, by means of two suitable diodes in series, it is possible to build

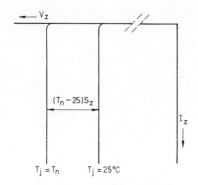

FIGURE 8. Typical dynamic characteristics of a voltage regulator showing the effect of a temperature change.

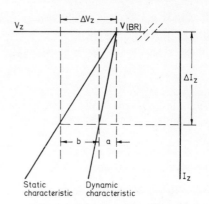

FIGURE 9. Derivation of static characteristics from dynamic characteristic.

stabilised supplies with a temperature coefficient that is nearly zero. In diodes with a breakdown voltage of about 5. V, the temperature coefficient can be zero at one specific current over a limited temperature range.

The static characteristic of the voltage regulator diode shows the variation of steady-state reference voltage with direct current. The change of reference voltage with direct current depends on (1) the dynamic resistance r_Z, which changes with current, (2) the rise in junction temperature because of the internal dissipation and the thermal resistance between the junction and ambient, and (3) the temperature coefficient S_Z.

When breakdown has occurred, the change in voltage is equal to a plus b as shown in Figure 9. Because the current before breakdown is very small, no sensible error is introduced by assuming that the characteristics coincide with the abscissa up to the breakdown voltage. Therefore, the total change in voltage ΔV_Z across the diode is given by

$$\Delta V_Z = a + b$$
$$= \Delta I_Z r_Z + V_Z I_Z R_{th(j-amb)} S_Z,$$

where V_Z is the voltage across the voltage regulator diode after breakdown, I_Z the current through the voltage regulator diode after breakdown, and $R_{th(j-amb)}$ the thermal resistance between junction and ambient.

The current before breakdown occurs is small; therefore there is no great error in saying that ΔI_Z equals I_Z. Hence,

$$\Delta V_Z = I_Z r_Z + V_Z I_Z R_{th(j-amb)} S_Z. \tag{3}$$

When the static characteristic cannot be obtained empirically, it can be obtained by plotting

$$V_Z = V_{(BR)} + \Delta V_Z$$
$$= V_{(BR)} + I_Z(r_Z + V_Z R_{th(j-amb)} S_Z) \qquad (4)$$

where $V_{(BR)}$ is the voltage at which breakdown occurs. Rearranging the equation gives,

$$V_Z = \frac{V_{(BR)} + I_Z r_Z}{1 - I_Z R_{th(j-amb)} S_Z}.$$

The slope R_Z of the static characteristic is given by

$$R_Z = \frac{V_Z - V_{(BR)}}{I_Z}.$$

Replacing V_Z by the expression in Equation (4) gives

$$R_Z = r_Z + V_Z R_{th(j-amb)} S_Z.$$

Sometimes the point of breakdown is not clearly defined, but the *specified* reference voltage V_{Zs} is quoted at a *specified* current I_{Zs} together with the dynamic resistance r_{Zs}. When these values of V_{Zs} and I_{Zs} are steady-state values the voltage V_Z at any other current I_Z can be determined from

$$V_Z = V_{Zs} + (I_Z - I_{Zs})(r_{Zs} + V_Z R_{th(j-amb)} S_Z).$$

Therefore,

$$V_Z = \frac{V_{Zs} + (I_{Zs})r_{Zs}}{1 - (I_Z - I_{Zs})R_{th(j-amb)} S_Z}.$$

The voltages and currents specified in the data for the voltage regulator diodes, however, are usually dynamic values that define points on the dynamic curve. When this is so, the increase in junction temperature is $V_Z I_Z R_{th(j-amb)}$ and not $V_Z(I_Z - I_{Zs})R_{th(j-amb)}$ because the heat generated at the pulsed current I_{Zs} is negligible. Therefore, when the values of V_{Zs} and I_{Zs} specified in the published data are pulsed values, the steady-state voltage V_Z at any steady-state current I_Z is given by

$$V_Z = V_{Zs} + (I_Z - I_{Zs})r_{Zs} + V_Z I_Z R_{th(j-amb)} S_Z,$$

that is

$$V_Z = \frac{V_{Zs} + (I_Z - I_{Zs})r_{Zs}}{1 - I_Z R_{th(j-amb)} S_Z};$$

this applies to 25 °C.   At any other temperature T

$$V_Z = \frac{V_{Zs} + (I_Z - I_Z)r_{Zs} + S_Z(T - 25)}{1 - I_Z R_{th(j-amb)} S_Z}.$$

Correction for Changes in r_Z

Because r_Z changes with current, the equations given above for V_Z are not strictly accurate, although they are good approximations.   When a higher degree of accuracy is required, a correction must be applied to the value of r_Z.

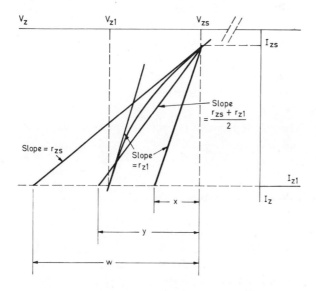

FIGURE 10.   Accurate determination of V_Z; curvatura of characteristic and errors greatly exaggerated.

When the current through a voltage regulator diode changes from I_{Zs} to I_{Z1}, the voltage across the diode changes to V_{Z1} as shown by the curve in Figure 10. However, if the value of r_Z were a constant and equal to the value at current I_{Zs}, the change in voltage caused by a current change of $I_{Z1} - I_{Zs}$ would be

$$(I_{Z1} - I_{Zs})r_{Zs} = w.$$

If the value of r_Z were a constant equal to the value at I_{Z1}, the change in voltage would be

$$(I_{Z1} - I_{Zs})r_{Z1} = x.$$

The value r_{Zs} gives too big a change, and the value r_{Z1} gives too small a change of voltage. A closer approximation to the true voltage is achieved if the change in voltage is taken as

$$y = \tfrac{1}{2}(w + x)$$

$$= \tfrac{1}{2}(I_{Z1} - I_{Zs})r_{Zs} + (I_{Z1} - I_{Zs})r_{Z1}$$

$$= \tfrac{1}{2}(I_{Z1} - I_{Zs})(r_{Zs} + r_{Z1}).$$

Values for the dynamic resistance at various currents can be obtained from curves in the published data. Therefore, when a more accurate value of V_Z is required, the value of r_Z in Equation (4) and the other expressions for V_Z should be replaced by the mean of the values of r_Z at I_{Zs} and at the new current.

Typical dynamic impedance (r_Z) curves plotted against Zener current at 25 °C for 400 mW diodes BZY88 are shown in Figure 11.

Breakdown Voltage Stabilisation Time

When a constant current is applied to a diode, thermal stability is not achieved immediately, but may be delayed by as much as six minutes. During this period the voltage across the diode changes because of the temperature dependence of the reference voltage. This change in voltage, however, is so small relative to the reference voltage that it is usually negligible.

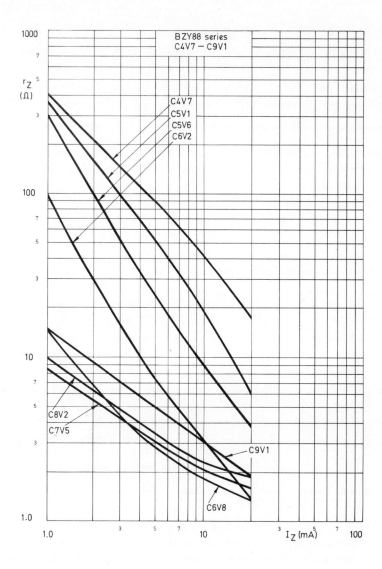

FIGURE 11. Typical dynamic impedance (r_Z) plotted against
Zener current.

Silicon Avalanche Diodes

Silicon avalanche diodes, otherwise known as controlled avalanche silicon rectifier diodes, are devices capable of dissipating energy in the reverse direction of the same order as in the forward direction.

Conventional rectifier diodes described previously are limited in their application to transient voltages within their peak reverse voltage rating. Transients above the breakdown rating will normally cause permanent damage or even total failure of the device. Therefore circuits had to be designed with sufficient safety margin to allow for any possible transient, or protection filters had to be incorporated.

The silicon avalanche diodes exhibit sharp turn-over characteristics, similar to the well-known voltage regulator (Zener) diodes (Refs. 17 to 19). The turn-over voltage of the controlled avalanche diodes ranges between 100 V to over 1 kV at various current ratings, depending upon the size of the device, up to about 250 A. They can be considered as an extension of the Zener diodes to higher voltages with relaxation on voltage tolerance and pulse duration limitation at the avalanche voltage.

Main applications of silicon avalanche diodes are in transient voltage suppression when circuits using transformers or inductive loads are switched. They are also used for high-voltage series applications.

As an example, avalanche breakdown voltage characteristics plotted against avalanche breakdown current of BYX30 series are shown in Figure 12.

The diodes are intended for use in high-frequency power supplies, thyristor inverters, choppers, ultrasonic systems, and multi-phase power rectifier applications up to 14 A mean forward current. The BYX30 diodes are capable of absorbing transient energy within the diode circuit without damage up to 9·5 kW for 10 μs square wave of 50 Hz pulse repetition frequency.

Transistor Switching Characteristics

The Transistor as a Switch

The transistor is basically a three-terminal device wherein the resistance between two terminals (collector and emitter) is either very low, viz. a fraction of an ohm or very high of the order of several megohms, depending on the voltage (or current) applied to the third terminal (the base) relative to the emitter. By way of an example consider Figure 13. In (a) is shown a p–n–p transistor with the appropriate voltage polarities applied to the

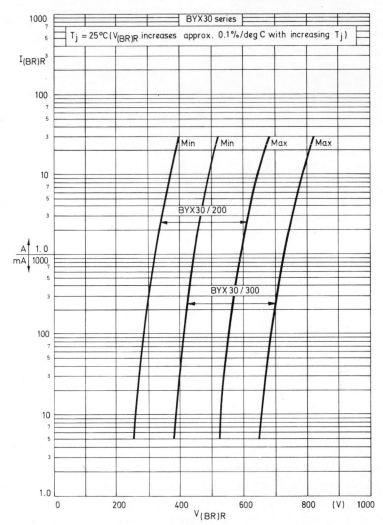

FIGURE 12. Avalanche breakdown voltage plotted against avalanche breakdown current.

collector and base terminals relative to the emitter which is at zero potential or earth. The corresponding conditions for an n–p–n transistor are shown in (b). If the p–n–p transistor circuit in (a) is considered, the following features are significant in that they are opposite to the corresponding features in the n–p–n transistor circuit in (b).

(1) The supply voltage to the collector load resistance (R_L) is negative relative to the emitter.

(2) The load current flows in the direction indicated by the arrow drawn against R_L, which direction is also indicated by the emitter arrowhead of the transistor symbol.

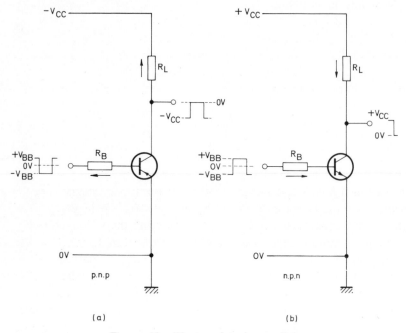

(a)   (b)

FIGURE 13.   The transistor as a switch.

(3) The base or signal current flows in the direction indicated by the arrow drawn against the base connection.   A property of the base terminal is that current can only flow in this one direction being virtually blocked off in the opposite direction.   Thus, when the base terminal is held at a negative potential $-V_{BB}$ relative to the emitter, base current flows, and the emitter-to-collector resistance is then at its low value so that the load current flows from emitter through collector.   The ratio of collector current (I_C) to the base current (I_B) is the current gain of the transistor and is expressed as

$$h_{FE} = \frac{I_C}{I_B};$$

h_{FE} is one of the transistor hybrid parameters signifying the forward current transfer characteristic for the grounded emitter circuit configuration shown in Figure 13. Further discussion on hybrid parameters as such is not necessary here as it is not relevant to the consideration of the transistor as a switch.

Normally in switching applications the base current supply is made more than adequate to allow the full collector load current to flow, i.e. $I_B > I_C/h_{FE}$. In this condition the transistor is said to be 'saturated'.

(4) When the base terminal is taken from a potential negative relative to the emitter, to a potential positive relative to the emitter, no current (except leakage) flows into the load R_L. In this condition the transistor is said to be 'cut off'.

Thus the analogy between transistor and mechanical/electromechanical switch is clear, where the mechanical toggle or armature is replaced by an electrical signal applied to the base of the transistor. The transistor is not as perfect a switch as mechanical contacts, however, since, when it is off, it has a much higher leakage and, when it is on, it has a higher resistance. Because of these limitations, power is dissipated in the transistor in both states. This constitutes a wastage, thus lowering the power efficiency of the circuit and, more important, it constitutes a limiting factor in the choice of a transistor which may have the necessary current and voltage ratings but not the power dissipation capability for a particular application. The rate at which heat generated internally owing to dissipation is dispersed through the structure of the transistor crystal and its surrounding envelope is a characteristic of the transistor known as thermal resistance, commonly designated by the symbol R_{th} and expressed as the internal temperature rise in degrees Celsius per watt. Nevertheless, the transistor has the advantages of small size, freedom from contact bounce, longer life, negligible maintenance, silent operation, and high-speed response.

Germanium Versus Silicon

In addition to the choice between p–n–p and n–p–n transistors there is also a choice to be made between the two most commonly used materials for transistor fabrication, viz. germanium and silicon. In spite of its higher cost, silicon offers the very attractive advantage of working at much higher internal temperatures than germanium (200 °C against 80 °C typically). This means that, size for size, transistors with the same thermal resistance characteristics can be used at higher power dissipation levels when made of silicon or, alternatively, can be used at the same dissipation but at higher ambient temperatures. Silicon transistors have leakage currents several orders lower than germanium and this normally means negligible dissipation in the off condition and greater thermal stability under high-voltage stress across collector and emitter. Unfortunately the saturation resistance levels obtainable with silicon tend generally to be higher and this means poorer efficiency in the on condition. Another significant feature of silicon transistors as compared with germanium is that the emitter–base junction voltage, i.e. the minimum voltage that must be applied before base current will flow to switch the transistor on, is higher for silicon, being typically 0·6 V as compared with about 0·2 V for germanium. Because of the higher base–emitter voltage required to turn a silicon transistor on and the lower leakage current of silicon, it is not usually necessary to reverse-bias the base–emitter junction to maintain the transistor cut-off as is the case with germanium. For reasons such as these germanium and silicon are not ordinarily compatible in a single design.

Ratings

Voltage

The transistor has two junctions, one between the collector and base terminals, and the other between the emitter and base terminals. The maximum reverse voltage that can be sustained across a semiconductor junction is a design characteristic of the junction. For example, alloy junction transistors generally have higher emitter–base breakdown voltages (typically a few tens of volts) than alloy diffused and silicon planar transistors which break down at only a few volts. This necessitates certain circuit modifications which will be discussed later. Specification of the collector–base and emitter–base breakdown voltages is therefore a simple matter of quoting the reverse breakdown voltages of the respective junctions. The specification of the collector–emitter breakdown voltage, however, is complicated by transistor action, i.e. it depends upon what condition the base terminal is held at relative to the emitter. Figure 14 illustrates that, with the base terminal open-circuited, the breakdown voltage is as indicated by $V_{(BR)CEO}$. With the base–emitter junction forward-biased, the breakdown voltage tends to be slightly lower as forward bias is increased. On the other hand, as the base–emitter junction is progressively reverse-biased, the breakdown voltage increases and approaches in the limit the value of the breakdown voltage of the collector–base junction. Once the junction has broken down, however, the voltage sustained across the collector to emitter falls back to the value $V_{(BR)CEO}$. Provided the collector current

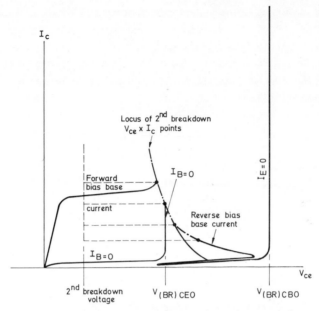

FIGURE 14.   Transistor breakdown voltage characteristics.

is limited so that excessive power dissipation does not destroy the junction, this voltage breakdown (*first breakdown*) is reversible, i.e. it does not of itself result in the destruction of the transistor.

Second breakdown is a term describing the collapse of the sustained voltage across collector to emitter to a level lower than the first-breakdown value $V_{(BR)CEO}$ (see Figure 14). The voltage at which second breakdown is initiated depends upon the bias conditions at the base terminal. As illustrated in Figure 14 by the dashed lines, the voltage at which second breakdown begins is lower as the forward-bias regions are entered, where the collector currents increase. In the reverse-bias region the voltages are higher and the collector currents lower. The reverse-bias condition proves to be more destructive to the transistor than the forward-bias condition, when second breakdown occurs. It is therefore evident that the application of heavy 'turn-off' drives to power transistors can be particularly destructive, if the simultaneous (V_{CE}, I_C) values during the switching-off transition, lie to the right-hand side of the locus of the second-breakdown points shown chain-dotted in Figure 14. With inductive collector loads the full supply voltage occurs across the transistor at relatively high load currents. Since the voltage across the load during the transient is a function of the rate of change of the collector current, special attention must be paid to the instantaneous (V_{CE}, I_C) values during switch-off, to ensure operation within the transistor manufacturer's safe limits. This information is usually given in the form of Figure 15. The causes and mechanism of second breakdown are extensively covered in the literature and need not be discussed here (Refs. 20 and 21).

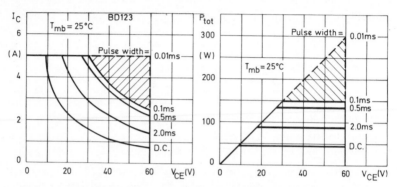

FIGURE 15.   Permissible working areas for d.c. and pulse operation. The shaded areas are for non-repetitive short-duration overload conditions only.

Power Dissipation

Apart from second-breakdown considerations the power dissipation in the transistor during the (V_{CE}, I_C) switching transient can be significant particularly for slow transistors switching inductive loads. Because of the thermal time constant of the crystal of the power transistor, high power dissipation transients of short duration relative to this time constant can be sustained without the junction temperature exceeding the maximum permissible value. The relevant data on this topic usually take the form of curves of effective thermal resistance (effective degrees Celsius per watt) . plotted against duration of the dissipation, with the duty cycle of the occurrence of the dissipation as a running parameter. Typical curves are shown in Figure 16. These curves show that the effective thermal resistance is lower when the duration of the transient dissipation is shorter and the duty cycle lower. When operating at high repetition rates, the average dissipation may be considerably increased by the recurrent transient dissipation. Thus the average or equivalent dissipation must take into account the dissipations during *off* state, the *saturated* state and the transients.   The maximum continuous power dissipation is calculated;

$$P_{C(max)} = \frac{T_{j(max)} - T_{amb(max)}}{R_{th(j-case)} + R_{th(h)}}$$

The value of the junction temperature $(T_{j(max)})$ used in the above expression should be less than the maximum junction temperature rating by a margin

$$(P_{C(transient)} \times R_{th(effective)}) \text{ degC.}$$

$T_{amb(max)}$ is the design maximum ambient temperature, $R_{th(j-case)}$ the thermal resistance junction to case (degC/W), and $R_{th(h)}$ the thermal resistance of the heatsink to the ambient (degC/W).

It must be noted that all transistor ratings applicable to a design should ideally be those appertaining to the maximum operating temperature of the junction.   The steady-state and pulse thermal conditions are discussed with more details in Chapter 2.

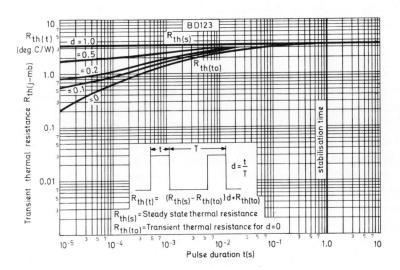

FIGURE 16.   Transient thermal resistance for various duty cycles plotted against pulse duration.

Current

The maximum current rating of a transistor is implied in the maximum power dissipation rating and in the maximum simultaneous (V_{CE}, I_C) values during switching transients. Nevertheless the current rating is always specified separately. The electrical characteristics of a transistor vary with the current passing through it and this factor may set limits to the useful current range over which a device may satisfactorily be used.

Characteristics

Leakage

The collector leakage current of a transistor depends upon the bias condition at the base. Figure 17 shows a typical (I_C, I_B) characteristic. In the forward-biased region the collector current is $h_{FE} \times I_B$. When $I_B = 0$, however, it will be observed that $I_C = I_{CEO}$ and, as the reverse-bias base current increases, I_C decreases to a minimum value I_{CBO} which is the collector–base leakage with no current flowing in the emitter. The relationship between I_{CEO} and I_{CBO} is

$$I_{CEO} = I_{CBO}(1 + h_{FE}).$$

This relationship may be understood simply by considering that the collector–base junction leakage current cannot flow through the base terminal since $I_B = 0$. Therefore it flows inside the transistor from base to emitter. The base–emitter junction is thus forward-biased and normal transistor action takes place so that the collector–base leakage current I_{CBO} is multiplied by the current gain h_{FE} of the transistor. In practice there is always a finite circuit resistance at the base terminal, as shown in Figure 17(a), so that the collector–base leakage current flowing through it generates a voltage which tends to forward-bias the transistor and turn it on when it is supposed to be off. The circuit resistance at the base terminal should therefore be low enough to ensure stable operation at high temperature. This is usually not a serious problem with silicon transistors, as has already been pointed out above.

Saturation Voltage

When a transistor is saturated, both the junctions are forward-biased and the junction voltages are consequently in opposition. The net junction

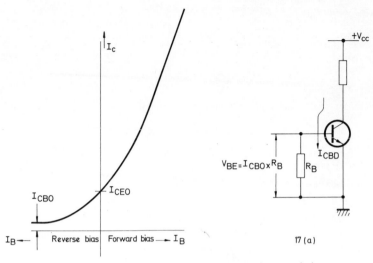

FIGURE 17. Transistor leakage current characteristic.

voltage appearing across the collector to emitter is one component of the saturation voltage. The other component is resistive and is indicated in Figure 18. The edge of the saturation regions is at zero voltage between collector and base, as seen in Figure 18, $V_B = V_C$. The saturation voltage $V_{CE(sat)}$ is usually specified at a particular value of collector current I_C and of base current I_B. Alternatively for high-current transistors the saturation resistance $R_{CE(sat)}$ may be specified.

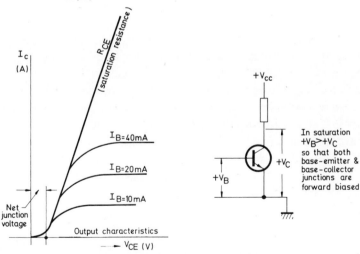

FIGURE 18. Transistor output characteristics.

Forward Current

Forward current gain (h_{FE}) is given by the relationship

$$h_{FE} = \frac{I_C}{I_B}$$

where I_C is the collector current and I_B is the base forward-bias current. This parameter varies with the value of I_C and, broadly speaking, it falls off at the high and low extremities of the collector current range.   The shape of the (h_{FE}, I_C) curves varies with transistor type, and typical curves are usually given in manufacturer's data.   The current gain h_{FE} also increases with the collector–emitter voltage, although this variation is not normally large as indicated by the slight upward slope of the output characteristics of Figure 18 in the active region of transistor operation, i.e. the region to the right of the R_{CE} line where the characteristics assume a linear aspect. The h_{FE} falls off quite markedly as the saturation resistance line is approached, as indicated by the downward slope of the characteristics in this region, where the transistor enters the saturation region of operation.   In practice, where power transistors are used as switches, the h_{FE} value of interest is that which indicates the lowest value of base current required to maintain the collector–emitter voltage at its lowest value for the particular collector current in question, i.e. on the saturation resistance line $R_{CE(sat)}$.

Switching Characteristics

Figure 19 shows the response of the collector current to a positive step of voltage applied through a resistance R_B to the base of an n–p–n transistor. There is a delay (t_d) after the application of the step at time $t = 0$ before the collector current begins to rise.   This is due to the fact that the emitter and collector depletion capacitances have a change of voltage across them of 0 to $+ V_j$ for silicon transistors which are not normally reverse-biased (see page 21), and of $-V_B$ to 0 for germanium transistors where V_j, the junction voltage, is small compared with $-V_{BB}$, before the transistor becomes forward-biased.   The time constant for this part of the response is approximately

$$(R_B + r_{bb'})(C_{TE} + C_{TC}) + R_L C_{TC},$$

where $r_{bb'}$ is the internal base resistance of the transistor, and C_{TE} and C_{TC}

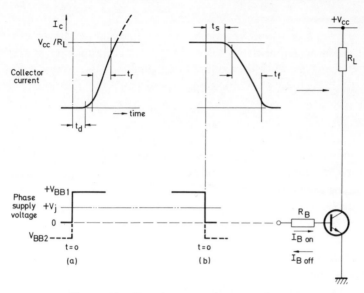

FIGURE 19. Transistor switching waveforms.

are the emitter and collector depletion capacitances respectively. The delay time is given by

$$t_d = \left\{ (R_B + r_{bb'})(C_{TE} + C_{TC}) + R_L C_{TC} \right\} \ln \left(\frac{V_{BB1} + V_{BB2}}{V_{BB1}} \right).$$

The rise of collector current according to the well-known theory (Refs. 22 to 24) of switching behaviour of transistors shows that the response in the normal active region is exponential and is expressed

$$i_C = I_C \left\{ 1 - \exp\left(-\frac{t}{\tau} \right) \right\}$$

where i_C is the instantaneous value of the collector current, I_C is the value of collector current aimed at, i.e. $h_{FE} \times I_B$ (the base current which equals $V_{BB1}/(R_B + r_{bb'})$ approximately), and τ is the collector current rise time constant and equals h_{FE}/ω_1. ($\omega_1 = 2\pi f_1$ where f_1 is the frequency at which the modulus of $h_{FE} = 1$.) From the above it becomes apparent that, although ω_1 is a small-signal parameter, its value may be determined from the rise time of the collector current of a transistor. The value of ω_1 so determined is $\bar{\omega}_1$. In practice the measured value of ω_1 is often at variance with the theoretical value. A more accurate value is obtained by extra-polation from a point on the gain h_{FE} versus frequency characteristic where the fall-off is asymptotic to 6 dB/octave. It is therefore common practice to use the gain–bandwidth product designated ω_T.

With a collector load resistance R_L, the total collector rise time constant is

$$h_{FE}\omega_T + C_{TC}R_L.$$

If a voltage drive is applied to the base of a transistor, i.e. $R_B = 0$, the collector current rise time constant is

$$\frac{h_{FE}}{\omega_T}\left(\frac{r_{bb'}}{r_{bb'} + h_{FE}r'_e} + C_{TC}R_L\right).$$

From this expression it is evident that the internal base and emitter resistances ultimately limit the drive current and, since r_e, the emitter resistance, varies inversely as the emitter current, this expression is not significantly different from the current drive time constant except at low current levels. Since the driving current and therefore the switching speed is ultimately limited by the internal base resistance, the factor $\omega_T/r_{bb'}$ is indicative of the relative switching speeds of transistors and is therefore a figure of merit; the higher the figure the faster the transistor.

Referring to Figure 19(a) the dashed portion of the collector current waveform indicates the current $h_{FE} \times I_B$ which would flow if the current were not limited by the circuit to the value of approximately V_{CC}/R_L. In practice switching transistors are overdriven in this way to achieve rapid switching speeds and to minimise dissipation in the conducting condition. Thus used, a transistor is said to be saturated. An approximate expression for rise time from 10% to 90% is

$$t_r = 0{\cdot}8\,\frac{I_C}{I_{B(on)}}\left(\frac{1}{\omega_T} + C_{TC}R_L\right).$$

When a saturated transistor is switched off, there is a delay or desaturation time t_s before the current begins to fall. Theoretical considerations give the following expression:

$$t_s = \tau_s \ln\left(\frac{I_{B(off)} + I_{B(on)}}{I_{B(off)} + I_C/h_{FE}}\right),$$

where τ_s is the desaturation time constant.

$I_{B(off)}$ is the reverse base current which can be driven though the base of a transistor during a switch-off transition of collector current by virtue of the minority carrier storage in the base region of the transistor. The heavier this turn-off drive current, the shorter will be the desaturation time and also the fall time t_f of the collector current. The turn-off transient of collector current follows the same law as the turn-on transient. In practice the type of construction of a transistor can have an overriding influence on the desaturation and fall times. Also second-breakdown considerations tend to put an embargo on heavy turn-off drive currents, and the disadvantage in conventional circuits of having to provide a reverse-bias voltage supply tends to limit the circuit designer's scope to attain fast turn-off times. However, high-ω_T transistors, which have high recombination rates of the minority carrier storage, enable a satisfactory turn-off performance to be obtained. Manufacturer's data usually provide information on turn-on and turn-off times and curves describing desaturation time variation with collector current and base turn-off current. Finally, it is worth noting that with slow switching transistors, i.e. low-frequency transistors, it is the parameter ω_T which has the dominant influence on switching performance, whereas with high-speed transistors i.e. high-frequency transistors, it is the depletion capacitances, circuit resistances, and stray capacitances which are more significant in limiting the switching speed. Other useful information concerning transistors can be found in Refs. 25 to 31.

Thyristor Characteristics

The thyristor or silicon controlled rectifier (SCR) is a four-layer semiconductor device having three electrodes as shown in Figure 20 (Refs. 32 to 35). The device has a reverse characteristic similar to that of an ordinary rectifier diode. Its forward characteristic (Figure 21) is such that, if no signal is applied to its gate terminal, the device will block positive anode-to-cathode voltages. However, by applying the correct signal to the gate the device can be made to conduct.

The gate circuit is dependent upon the anode circuit in one respect —a thyristor will never trigger if the anode circuit limits the anode current to below a certain value known as the latching or pick-up current I_L. Similarly, conduction will cease when the anode current is reduced to below the

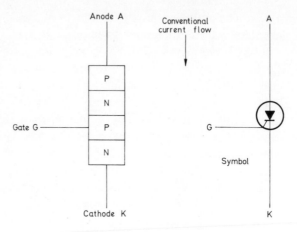

FIGURE 20.   The thyristor.

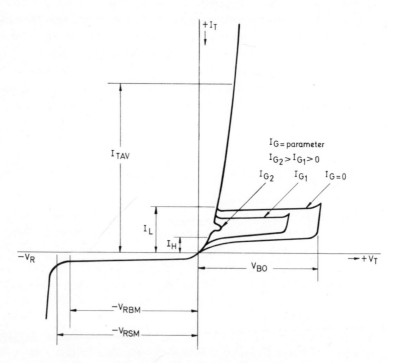

FIGURE 21.   Typical thyristor anode characteristic with I_G as a parameter in the first quadrant.

holding current I_H. The reduction in current can be effected by reducing the anode voltage to zero or to a negative potential.

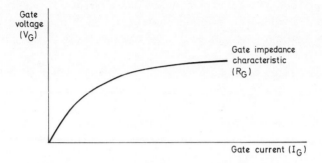

FIGURE 22. Typical gate impedance characteristic.

The p–n junction between gate and cathode terminals behaves as a diode in series with a resistance, and triggering makes the diode conduct in the forward direction. Figure 22 shows a typical gate–cathode impedance characteristic. The gate–cathode impedance varies with temperature and with different thyristors of the same type. Therefore in the manufacturer's data maximum and minimum gate impedance characteristics are given for each type of thyristor in the form of a graph shown in Figure 23. This spread of the characteristics must be taken into account when triggering circuits are designed.

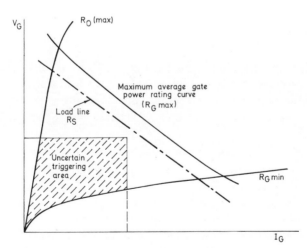

FIGURE 23. Spread of the gate impedance characteristics for a BTY91-200R.

If a positive voltage is applied between the gate and cathode of a typical thyristor, the gate current will increase, as shown in Figure 22, and at some point along the R_G curve the thyristor will be triggered. This point is largely independent of voltage on the anode, but, as previously mentioned, the value of gate current and gate voltage will vary somewhat from one thyristor of the same type to another, and also for different temperatures.

For a given gate voltage, the required gate current will be large for lower temperatures. For example, to trigger a typical thyristor such as the Mullard BTY91-200R with a gate voltage of 3 V, the gate current required at + 125 °C is 25 mA, at + 25 °C is 42 mA, and at − 55 °C is 80 mA.

Thus there is a shaded area (Figure 23) bounded by the maximum and minimum gate characteristics and the minimum voltage and current necessary to trigger all thyristors of a given type which is called the circuit triggering area. A properly designed triggering circuit should trigger all controlled rectifiers of a given type. Therefore the gate signal source should produce voltages and currents in excess of the values shown by the shaded area, but the gate signal must not exceed the maximum gate power rating, which for this particular thyristor is 500 mW.

D.c. or a.c. triggering can be applied to the gate. If a.c. triggering is used, however, the reverse gate voltage must be suppressed to a value below the maximum reverse gate voltage rating. A diode between gate and cathode can be used for this purpose. More information on thyristor characteristics and triggering can be found in the Refs. 36 to 43.

Characteristics

Breakover voltage (V_{BO}). The minimum forward breakover voltage is a minimum value of positive anode voltage at which a thyristor switches into the conductive state with the gate open-circuited.

Forward voltage (V_T) is the voltage between anode and cathode at a specified anode current.

Reverse voltage (V_R) is the reverse voltage between anode and cathode, V_{RWM} the crest working reverse voltage, and V_{RRM} the repetitive peak reverse voltage (1% duty cycle at 50 Hz).

Forward current (I_T) is the average value of anode current, $I_{T(AV)}$ the maximum mean on-state current, I_L the latching current (the value of I_T at which thyristor remains on with gate signal removed), and I_H the holding current (the value of I_T below which thyristor returns to blocking state with the gate open-circuited).

Other Devices

A number of other semiconductor devices exist which are used in conjunction with power supplies and inverters.   The devices will be briefly described with reference to their main characteristics.

Tunnel Diodes

A typical (V, I) characteristic curve of a tunnel diode is shown in Figure 24. As the voltage applied across the device is increased gradually, the diode current increases rapidly.   The peak in the characteristic is followed by a negative resistance region during which the diode current decreases until the valley is reached after which the current increases again owing to the normal diode conduction (Refs. 44 to 58).

The most common values of the peak current I_p lie between 1 and 20 mA for small-signal applications.   For power switching, devices are available with peak currents of up to 250 A.

The valley current I_v is usually small compared with the peak current I_p. Normally I_v is less than $(1/10)I_p$.   The voltage V_p at the peak current may be between 100 and 150 mV or higher, depending on the value of the peak current and whether the device is made of germanium or gallium arsenide. The voltage at the valley current varies between 250 and 350 mV for germanium and between 500 and 700 mV for gallium arsenide tunnel diodes.

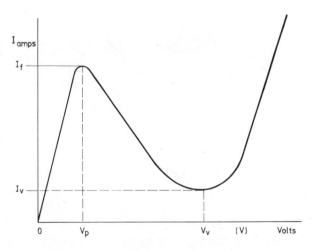

FIGURE 24.   Tunnel diode characteristic.

Four-layer Diodes

The four-layer diode is a semiconductor device similar to a thyristor except that it is a two-terminal device only. The diode behaves as a voltage-controlled switch having two stable states. The *off* state which has characteristics similar to a reverse-biased diode is shown in Figure 25. The off-state resistance of the diode is high, of the order of a megohm or more, and therefore only a small leakage current can flow.

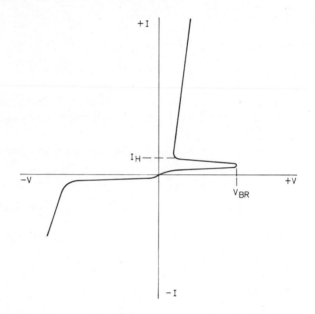

FIGURE 25. Typical four-layer diode characteristic.

In the *on* state the series resistance of the device is low, of the order of several ohms, allowing high current to flow, which has to be limited by an external resistance.

The forward characteristic of the four-layer diode as seen from Figure 25 is similar to the thyristor characteristic. If a voltage higher or equal to the breakover voltage is applied across the device, the diode switches to its on state and remains there as long as the current flowing through the device is maintained at a value higher than the holding current I_H. When the diode current falls below the value of I_H, the device switches back to its off condition.

More information on four-layer diodes can be found in Refs. 39, 59 to 63.

Unijunction Transistors

The unijunction transistor (known previously as the double-base diode) is a three-terminal device.   The three terminals are the emitter connected to the p-side of the junction, and base one and base two connected to either end of the n-type silicon bar forming the base structure (Refs. 33, 39, 63 to 70).   The device has only one p–n junction and therefore is different from the conventional transistor.   The unijunction transistor symbol, together with nomenclature and an equivalent circuit, are shown in Figure 26. The equivalent circuit is valid for emitter currents I_E equal to or less than the peak point current I_p.

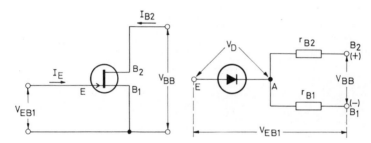

FIGURE 26.   Unijunction transistor and its equivalent circuit. [By courtesy of Motorola Inc., Ref. 64.]

When a positive voltage V_{BB} is applied across the base terminals B_2 and B_1 a current that flows in the interbase structure is given by

$$I_{B2} = \frac{V_{BB}}{r_{BB}},$$

where r_{BB} is the interbase resistance such that

$$r_{BB} = r_{B1} + r_{B2}.$$

A static characteristic curve for a single value of V_{BB} is shown in Figure 27. The peak point voltage is defined by the following equation;

$$V_p = V_D + nV_{BB}.$$

where V_D is the forward voltage drop of the p–n junction and n is the intrinsic stand-off ratio given by

$$n = \frac{r_{B1}}{r_{BB}}.$$

As seen from the emitter characteristic, the unijunction transistor exhibits a clearly defined stable negative resistance region. Since the negative resistance region is current-stable, simple relaxation oscillators and bistable circuits may be designed without the use of inductances.

Typical applications include triggering circuits for thyristors, sawtooth oscillators, pulse generators, and frequency dividers.

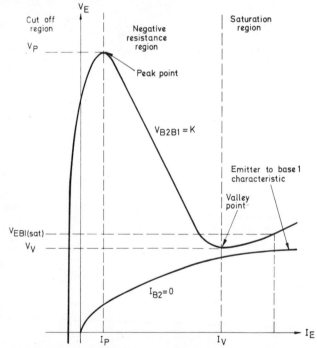

FIGURE 27.   Emitter characteristic curves of unijunction transistor. [By courtesy of Motorola Inc., Ref. 64.]

SCS's, Diacs, and Triacs

Three other semiconductor devices which may be of interest to the reader are listed here with references.   These are silicon controlled switches (SCS's) (Refs. 33, 39, 71, 72), diacs (Refs. 33, 73, 74), and triacs (Refs. 33, 75 to 77).

CHAPTER 2. THERMAL CONSIDERATIONS
OF SEMICONDUCTOR DEVICES

The junctions of semiconductor diodes or transistors must not exceed their maximum temperature ratings. The temperature attained is determined mainly by the electrical dissipation at the junctions and by the ability of the device to transfer heat to its surface, to any auxiliary dissipators (flags, fins, or heatsinks), and thence to ambient. Thus the thermal resistance of all the conductive, convective, and radiative heat paths must be known.

To protect the junction of a semiconductor device, manufacturers usually stipulate a maximum junction temperature for that device. Under pulsed current conditions the thermal capacitances of the whole semiconductor device tend to limit the temperature changes at the junction. Thus the relevant data for calculating steady-state junction temperatures introduce unnecessarily large safety factors if used for pulses.

The calculation of thermal resistances, power ratings, and the design of dissipators are discussed for steady-state and pulse conditions.

Steady-state Conditions

For safe and efficient operation of semiconductor devices, it is essential that the maximum rated junction temperature is not exceeded. Therefore thermal resistance—the dependence of junction temperature on device dissipation—is one of the parameters always given in published data, and its measurement is of considerable importance (Ref. 78).

Definitions

The thermal resistance of semiconductor devices is defined (Ref. 79) as follows:

'*Thermal resistance, effective (of a semiconductor device).* The effective temperature rise per unit power dissipation of a designated junction above the temperature of a stated external reference point under conditions of thermal equilibrium.

Under steady-state conditions, the thermal resistance of any thermal conductor is the ratio of the temperature drop across the conductor to the heat transfer rate through it.

When a continuous power is dissipated in a semiconductor device, the heat generated will cause the junction temperature to rise until the temperature difference between junction and ambient (that is the final cooling medium) is such that a stable thermal state exists. In this state, the heat transference from the device is equal to the heat generated in the device.

Although the heat is transferred by conduction, convection, and radiation, in general the overall heat transference from the junction to the surroundings should be considered. As the expression for current transfer, $I = E/R$, is comparable with that for heat transfer, an electrical analogue may be used in which current represents heat flow, voltage represents temperature difference, and resistance represents the opposition to the flow of heat energy—that is the thermal resistance.

Units and Symbols

The unit of heat transfer is joules per second (watts). The unit of thermal resistance may be either degrees Celsius per watt or degrees Celsius per milliwatt, the choice of unit depending on the power-handling capacity of the device.

The British Standards symbol for thermal resistance (Ref. 80) is R_{th}, used with suitable subscripts. Other symbols have been used, but they are not recommended.

The symbol R_{th} with no subscript is generally used to denote the total or overall thermal resistance. If an additional single subscript is used to denote a particular component of the overall thermal resistance, either an equivalent thermal resistance circuit or a clear definition should be given. However, in many cases the use of two subscripts may be sufficient to make the relevant thermal resistance clear at sight. Typical subscripts are as follows:

j, junction;   a, ambient;
c, case (or e, envelope);   m or mb, mounting base;
i, infinite heatsink;   h or hs, heatsink.

Thus $R_{th(j-mb)}$ is the thermal resistance from junction to mounting base.

Thermal Analogue

It is generally assumed that the heat transfer is by one main path, from the junction to the envelope of the device and then from the envelope to the ambient. There is an alternative path via the lead-out wires, but in most instances the simple approximation is still valid, as this is a minor heat path.

With power transistors, where the crystal is directly connected to the header, typical internal thermal resistance values of from 7·5 degC/W to < 1 degC/W are obtained. But, when the envelope of the device is the sole heat dissipator, the external thermal resistance is so large in proportion that the overall value is approximately equal to the external value. For example, for the OC28 in TO-3 envelope,

$$R_{th(c-a)} = 35 \text{ degC/W and } R_{th(j-c)} = 1 \cdot 5 \text{ degC/W.}$$

The corresponding equivalent circuit is shown in Figure 28(a).

Beyond a limit of about 300 to 400 mW it becomes impracticable to increase the size of the base and the thickness of the leads to reduce the external thermal resistance to a value comparable with the internal thermal resistance; therefore power devices are specifically designed for mounting on a heatsink. The effect of a heatsink or dissipator is to increase the effective heat-dissipation area. On the equivalent circuit, the effect is to produce an alternative low thermal resistance path to the heat flow from the case to the ambient, shown as $R_{th(3)} + R_{th(4)}$ in Figure 28(b), where $R_{th(3)}$ is the thermal resistance between the mounting base and heatsink by conduction (typically 0·2 to 3 degC/W) and $R_{th(4)}$ is the thermal resistance from heatsink to ambient by radiation and convection (typically 1 to 10 degC/W).

In order to utilise the full power-handling capacity of a device as indicated by $R_{th(j-c)}$, there must be no temperature rise between case and ambient. Such a condition occurs only when the external thermal resistance $R_{th(c-a)}$ is zero; that is when the device is mounted on a hypothetical infinite heatsink. The concept of an infinite heatsink is used when defining the internal thermal resistance between junction and case; in practice it can never be realised.

The greater the ratio of $R_{th(j-c)}$ to $R_{th(c-a)}$, the closer is the approximation to an infinite heatsink, and the power-handling capacity of the device approaches the limit defined by $R_{th(j-c)}$. With power devices, where $R_{th(j-c)}$ is low, very large heatsinks are required to obtain low enough values of $R_{th(c-a)}$ to utilise the full power-handling capacity. At lower dissipation

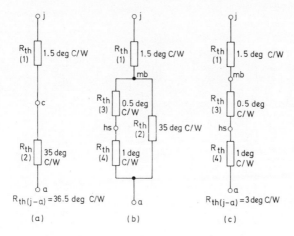

FIGURE 28.   Equivalent circuits for high-power transistors.

levels, smaller heatsinks (1 to 10 degC/W) may be safely tolerated with a consequent saving of space, unless the temperature rise is to be restricted to improve the stability of temperature-sensitive parameters.

When a heatsink is used, the radiation and convection from the case are generally small enough to be ignored (that is $R_{th(2)} > R_{th(3)} + R_{th(4)}$), and the circuit simplifies to that shown in Figure 28(c).

Heatsinks

Any form of plate, flag, or cylinder that increases the effective heat-dissipating area of a device, and hence the radiation and convection losses, will reduce the external component of thermal resistance.   Generally, when the published data sheet of a low-power device gives two values of thermal resistance—the overall value ($R_{th(j-a)}$) and the internal value ($R_{th(j-c)}$)—a comparison of the two figures will show that the external value constitutes a significant proportion of the whole.   In such instances a considerable improvement in power-handling capacity may be obtained by the use of a dissipator.

For metal envelopes of the TO-5, TO-9, and other types, the external value of thermal resistance is considerable, say, 210 degC/W.   Commercial forms of finned cylinder or flag dissipator, for use in free air, have a typical value of 70 degC/W.   The use of such a dissipator can reduce the value $R_{th(c-a)}$ to 52 degC/W, as shown in Figure 29.   If the internal thermal

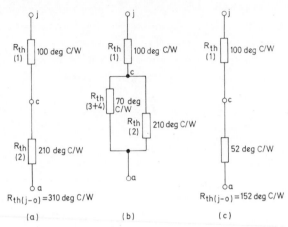

FIGURE 29. Equivalent circuits for low-power
transistors.

resistance of the device is comparatively low (< 100 degC/W), even $R_{th(c-a)}$ = 50 degC/W is still significant, and a considerable increase in power-handling capacity may be obtained with a larger heatsink.

Both flag and cylindrical finned types generally rely on a tight friction fit between the metal envelope and the heatsink for good conductive heat transfer.   Some forms of flag or finned cylinder are designed for mounting, if required, onto a larger heatsink.   Care should be taken to ensure a good thermal connection to the main heatsink, as a value of contact thermal resistance of, say, 10 degC/W will largely nullify the advantage offered by the larger surface area.

Considerable work has been done in the field of heatsinks for power devices, and commercial forms of heatsink are available in the range 1 degC/W to 3 degC/W.   These heatsinks are generally of the finned variety, where the necessarily large surface area is obtained in a comparatively limited space.   Where an excessively large heatsink is to be avoided, the following general points should be noted.

Heat is dissipated from the heatsink by convection and radiation.   In both cases the surface area, although it is important, is not the only factor that must be considered.   The temperature difference, the conductivity, thickness, and emissivity of the material, the shape and orientation of the heatsink, and the type of convection all play their part (Refs. 11, 81 to 89).

To avoid an excessive temperature gradient in the heatsink itself, a high-conductivity material should be used.   Copper and aluminium, with conductivities of 3·8 and 2·3 W/cm degC respectively, are the most usual materials.   For the same surface conditions an aluminium heatsink should be 3·8–2·3 times (respectively) thicker than the corresponding copper heatsink to obtain the same value of thermal resistance.

Radiation losses may be increased by improving the surface.   Best results are obtained with matt black finishes, typical emissivities being of the order of 0·9.

For free-air conditions, vertical mounting gives improved convective losses.   As a rough guide, 25% to 30% more area is required if the plate is horizontally mounted.   When air is free to move over both faces, the total surface area is effective.   The use of a chassis as a heatsink, where the air flow is restricted to one side, gives a reduction of the effective area by approximately one-half.

For calculation of losses, the area considered is the total area.   However, ·with a plane heatsink (for example, 25 cm × 25 cm) the size of the heatsink may be defined either by the total area (1250 cm^2) or by the area of one face (625 cm^2).

If a plane, vertical, blackened copper heatsink 3 mm thick, in free air, is taken as a standard, the following table gives a rough pessimistic guide to the total area required.

R_{th} (degC/W)	10	5	2	1
Total area (cm^2)	62·5	250	625	1250

At high ambient temperatures, in restricted spaces, or for values of $R_{th} \leqslant 1$ degC/W, forced air convective cooling is often used to reduce the effective thermal resistance of a heatsink.

The most practical way of obtaining a nearly 'infinite' heatsink is by liquid convection cooling.   With transformer oil as coolant, a typical figure of 10 degC/W for every 6·45 cm^2 of heatsink area may be obtained.

In all critical cases, the calculated value should only be taken as a guide, and the actual value of thermal resistance should be practically verified. This is generally done by monitoring the mounting base or stud temperature with a thermocouple.

Thermal Resistance Mounting Base to Heatsink

A thermal resistance $R_{th(3)}$ is shown in Figure 28, between the mounting base and the heatsink of the power transistor.   There are two reasons for this.

(i) When a single heatsink is used for several devices, some form of electrical insulation is often necessary. Unfortunately, good electrical insulators are generally poor thermal conductors; so the insertion of an insulating medium may considerably increase the thermal resistance between mounting base and heatsink.

The most common method of insulating the device is by means of a mica or Teflon washer. Mica has a lower electrical resistance and dielectric strength than Teflon, but a considerably lower thermal resistivity. Washer thicknesses vary from 0·025 to 0·15 mm, being governed by both mechanical and electrical considerations.

An alternative method is to deposit a thin insulating layer on either the device or the heatsink by means of anodising or similar techniques. It is desirable to check electrically that no short-circuit exists.

(ii) When trying to simulate an infinite heatsink or to obtain a very low external R_{th}, the thermal resistance of the insulation (0·2 degC/W to 3 degC/W) may be too great to be tolerated. Individual heatsinks requiring no insulation are then necessary. Even without an insulating layer, some thermal resistance exists because of the presence of a thin layer of air between the two surfaces. This is the contact or interface thermal resistance.

For low contact resistance, the following conditions are required.

(1) Sufficient pressure. Generally, a specified torque is given for stud-mounted devices, for example, see the following table for Unified Screw Threads:

Thread	Torque (lb in)	Torque (kg cm)
No. 10-32 UNF	15	17·5
$\frac{1}{4}$-28 UNF	30	35
$\frac{1}{2}$-20 UNF	150	175

(2) A plane surface to the heatsink. This should be machined or punched and free from burrs.

(3) A plane mounting base on the device. This is the manufacturer's responsibility. Tolerances of $0°$ to $0°\ 20'$ and $0°$ to $0°\ 30'$ are typical.

A common method of reducing the interface thermal resistance produced by the air layer is to substitute an alternative medium with a lower thermal resistance, such as a thin film of oil or grease. Silicon compounds which have a higher conductivity, which are non-corrosive to metals, and which are impervious to moisture are often used. A reduction in the interface thermal resistance of from 20% to 40% is often possible by this means.

When two poor mating surfaces give a high interface thermal resistance, an improvement may be achieved by using a thin lead or solder washer to 'pick up' any irregularities. The thickness of the washer may be anything between 0·1 to 0·2 mm.

Relative Importance of Internal Thermal Resistance in Determining Power-handling Capacity

From the I.R.E. definition, the general formula for thermal resistance is

$$R_{th(s-r)} = \frac{T_s - T_r}{P_s},$$

(6)

where T is the temperature in degrees Celsius, P is the power, and s and r are the source and reference, respectively. Therefore by simple transposition of the general formula

$$P_s = \frac{T_s - T_r}{R_{th(s-r)}}.$$

(7)

A high power-handling capacity may be obtained by the use of a device having a low internal thermal resistance mounted on a large heatsink, as previously discussed. However, there are two other possible means of obtaining a high value of maximum dissipation, both of which increase the numerator of the right-hand side of Equation (7).

(i) *Low reference temperature* T_r. If the ambient temperature T_a is taken as reference, the permissible temperature rise $T_s - T_r$ may be increased, and consequently P_s increased, by reducing T_a. In equipment design this would entail adequate ventilation, possibly forced convention cooling, and, where possible, isolation of the device from major heat sources.

(ii) *High source temperature* T_s. The alternative method is to increase T_s, that is to select a device with a high $T_{j(max)}$. The permitted value to $T_{j(max)}$ for silicon devices is from 150 °C to 200 °C, whilst for germanium devices it is only 75 °C to 100 °C. Therefore, for given values of $R_{th(s-r)}$ and T_r, the maximum dissipation for a silicon device is far greater than for a germanium device, the more so for a device whose $T_{j(max)}$ is 200 °C compared with another whose $T_{j(max)}$ is 150 °C.

Significance of $R_{th(j\text{-}c)}$

Consider two devices; A with $R_{th(j-c)} = 1$ degC/W, and B with $R_{th(j-c)}$ = 2 degC/W. Only when mounted on infinite heatsinks, all other things being equal, will device A have double the power rating of device B. In those instances where a low external thermal resistance is unnecessary, the internal value may give very little indication of the relative power-handling capacity of the devices.

For example, device A with $R_{th(j-c)} = 1$ degC/W, mounted on a heatsink such that $R_{th(c-a)} = 5$ degC/W (that is with an overall thermal resistance of 6 degC/W) will, all other things being equal, have a smaller power-handling capacity than device B mounted on a heatsink such that $R_{th(c-a)} = 2$ degC/W (that is with an overall thermal resistance of 4 degC/W).

Device B, even with the same value of $R_{th(c-a)}$ (2 degC/W), provided that it has a higher permissible $T_{j(max)}$, may have a greater maximum power handling capacity than device A.

Thus, let the reference temperature be 60 °C. If $T_{j(max)}$ of device A is 75 °C and $T_{j(max)}$ of device B is 90 °C, then from Equation (7)

for device A
$$P_s = \frac{75 - 60}{1 + 2} = 5 \text{ W},$$

and for device B
$$P_s = \frac{90 - 60}{2 + 2} = 7\cdot5 \text{ W}.$$

Heatsink Design

Semiconductor devices, which include rectifier diodes, power transistors, and thyristors, when operated at high currents may dissipate appreciable amounts of power due to the forward voltage drop across them. In addition, the reverse, transient, and switching dissipation need also to be taken into account especially when a high frequency of operation is encountered.

To ensure that the maximum mounting base temperature is not exceeded when a device is used at high powers, it is necessary to apply some form of cooling to conduct the heat away from the junction. This is achieved by bolting down the device to a convection cooled heatsink.

Sometimes forced air cooling is used to reduce the size of the heatsink or to increase its effectiveness. Water or oil cooling may be used when kilowatts of power are handled, but this is outside the scope of this book.

Most rectifier diodes and thyristors are supplied with stud mounting, whereas most power transistors are supplied with flat cases with holes for screwing down to a heatsink. The heatsink or cooling fins are usually supplied by the user rather than by the manufacturer of the device.

Such devices are known as case-rated, that is the rating is given in terms of the temperature on the case of the device.

Many high-power rectifiers and thyristors are supplied as units or stacks for bridge operation in single-phase or three-phase supplies. These units are complete with cooling fins and are rated with reference to a certain ambient temperature.

Although heatsink design is similar for all semiconductor power devices, the devices differ in the way their rating is given. In the case of rectifier diodes and thyristors the rating is in terms of average forward current, whereas transistors are rated in terms of power. The manufacturers therefore provide additional information for rectifier diodes and thyristors in the form of a graph.

As an example, curves relating total power dissipation to the maximum mounting base and ambient temperatures for various values of mean forward current and heatsink thermal resistance for the BYX38 diodes (Ref. 9) are shown in Figure 30.

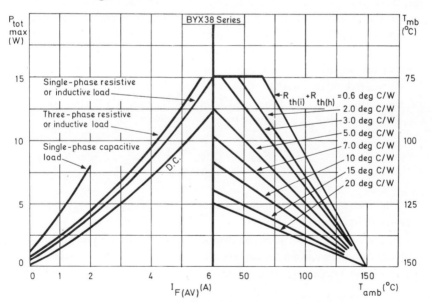

FIGURE 30.   Maximum mounting base and ambient temperatures for various values of mean forward current and heatsink thermal resistance.

The method of using the curves is as follows. Starting with the curve of maximum dissipation as a function of average forward current, for a particular current value trace upwards to meet the appropriate curve. Then, trace horizontally across until the line reaches the appropriate $R_{th(i)}$ + $R_{th(h)}$ curve. Finally trace downwards to determine the maximum ambient temperature.

The various components of the rise of junction temperature above ambient are illustrated below.

$R_{th(j\text{-}mb)} = 5\cdot0 \text{ degC/W}$ $\left\{\begin{array}{l} \textit{junction temperature} \\[1em] \textit{mounting base temperature} \end{array}\right.$

$R_{th(i)} = 0\cdot6 \text{ degC/W}$ $\left\{\begin{array}{l} \\ \textit{heatsink temperature} \end{array}\right.$

$R_{th(h)}$ $\left\{\begin{array}{l} \\ \textit{ambient temperature} \end{array}\right.$

where $R_{th(j\text{-}mb)}$ is the thermal resistance from the diode junction to its mounting base, $R_{th(i)}$ is the contact thermal resistance for the maximum torque as specified in the data, and $R_{th(h)}$ is the thermal resistance of the heatsink.

The thermal resistance of the heatsink depends on the cooling conditions under which the rectifier diode is used and also on the dimensions, position, and surface conditions of the heatsink.

Further data to help the designer consist of a graph showing values of $R_{th(h)}$ for a blackened, vertical heatsink (Figure 31).

The value of $R_{th(h)}$ is given by

$$R_{th(h)} = \frac{T_{mb} - T_{amb}}{P_{tot(max)}} - R_{th(i)}. \qquad (8)$$

The above method of using the curves applies both to rectifier diodes and to thyristors.

In the case of transistors the ratings will normally be that of the power and the maximum junction temperature so that the value of the heatsink required can be calculated from the equation

$$R_{th(h)} = \frac{T_j - T_{amb}}{P_{tot(max)}} - R_{th(i)}. \qquad (9)$$

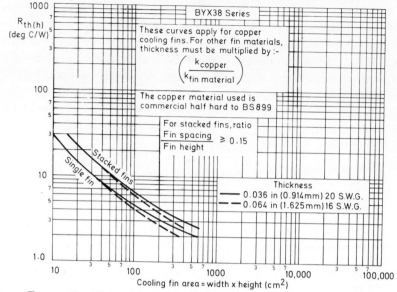

FIGURE 31. Thermal resistance of blackened, vertical, square cooling fin when used in free air.

Steady- state Power Rating

The maximum power that a device can dissipate is given by

$$P_{s(max)} = \frac{T_{j(max)} - T_{amb}}{R_{th(j-amb)}}, \qquad (10)$$

where $P_{s(max)}$ is the maximum permissible steady-state dissipation, $T_{j(max)}$ the maximum permissible operating temperature of the junction, T_{amb} the ambient temperature, and $R_{th(j-amb)}$ the total thermal resistance between junction and ambient. The value of $R_{th(j-amb)}$ quoted in the published data is for use only with devices that have no heatsinks and are mounted in free air. With other devices, $R_{th(j-amb)}$ depends on the type and size of the heatsink used, but can be determined from

$$R_{th(j-amb)} = R_{th(j-mb)} + R_{th(i)} + R_{th(h)}, \qquad (11)$$

where $R_{th(j-mb)}$ is the thermal resistance between junction and mounting base, $R_{th(i)}$ the thermal resistance between mounting base and heatsink, and $R_{th(h)}$ the thermal resistance between heatsink and ambient.

As an example how Equation (10) is used, consider one of the BZY91 series of voltage regulator diodes. This series has a maximum power rating of 75 W and nominal breakdown voltages that range from 10 to 75 V. If the diode has to dissipate only 40 W, what is the maximum permissible ambient temperature when the thermal resistance from the heatsink to ambient is 1·5 degC/W? (The value of $R_{th(h)}$ for different types and heatsinks can be determined from graphs printed in the published data.) For the BZY91, $R_{th(j-mb)}$ is 1·47 degC/W and $R_{th(i)}$ is 0·2 degC/W. Therefore,

$$R_{th(j-amb)} = 1·47 + 0·2 + 1·5$$
$$= 3·17 \text{ degC/W}.$$

The maximum permissible junction temperature for the BZY91 is 175 °C. Therefore,

$$40 = \frac{175 - T_{amb}}{3·17}$$

and

$$T_{amb} = 48·2 \text{ °C}.$$

Hence, under the conditions specified, the BZY91 can be used with an ambient temperature not exceeding 48 °C.

Pulse Conditions

When the power-handling capacity of a semiconductor device is estimated by the manufacturer before the relevant data are published, the fundamental restriction which is observed is that of the temperature of the semiconductor junction. Exceeding this maximum temperature rating could lead to damage within the device, and sometimes even to immediate catastrophic failure. To help circuit designers observe this absolute maximum rating, without having to measure the junction temperature itself, the published data have provided information which can be used to calculate conditions under which no damage will occur. The relevant information given includes an absolute maximum power and junction temperature, a value of the steady-state thermal resistance, and an equation relating total power to the maximum junction temperature and to either the case temperature or the ambient temperature. Under pulsed conditions, however, it is possible for the peak power to exceed the steady-state maximum, provided that the peak does not last long enough to cause the junction temperature to exceed its rating. Therefore, the information given to safeguard the device under pulsed conditions includes a maximum averaging time for the current rating or pulse rating charts. These charts are for establishing suitably modified values of thermal resistance for calculations based on various pulse conditions. The derivation of these modified values of thermal resistance is the topic of this present section (Ref. 90). This section is complementary to the previous section on steady-state thermal resistance and the procedures for examining junction temperature in both.

Reference to Steady- state Thermal Resistance

When there is steady dissipation, in which dynamic equilibrium of heat flow from the junction to the ambient is reached, the thermal path may be considered to be a simple network of thermal resistances. In these circumstances the junction temperature is given by the simple expression

$$T_j = T_{amb} + PR_{th} \qquad (12)$$

where T_j is the junction temperature, T_{amb} the ambient temperature, P the steady-state dissipation, and R_{th} the steady-state thermal resistance between junction and ambient.

However, thermal capacitance is associated with the several regions of the device. Thus the junction temperature response following a power step is a function of time. In particular, the maximum temperature reached under pulsed load conditions is lower than for the same power level operating continuously, if the pulse duration is less than 'the stabilisation time'. Hence, to produce an expression comparable with Equation (12) above and applicable to these conditions, the concept of transient thermal resistance, which is also a function of time, is used.

Concept of 'Transient' Thermal Resistance

When a step function of power is applied to a device, the response of its junction temperature is of the general form shown in Figure 32. For this single-step function the concept of 'transient' thermal resistance (designated $R_{th(t,0)}$) may be used to correlate the junction temperature at any instant and the magnitude of the power step. The value of $R_{th(t,0)}$ at this instant is defined as the ratio of the change of temperature at the time t after the power step to the change in power. Thus $R_{th(t, 0)}$ is a monotonic parameter, increasing with time in exactly the same manner as the heating curve. The single power step is rarely encountered in practice, but the response to this step will be shown later to be of considerable importance.

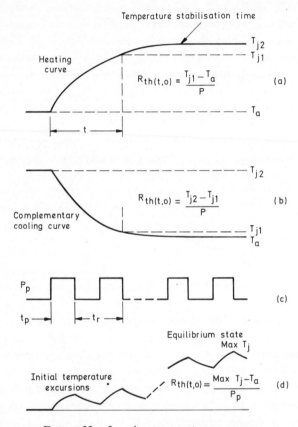

FIGURE 32.   Junction temperature response.

The concept of transient thermal resistance can be extended to embrace the more usual case of a train of rectangular power pulses, and the symbol $R_{th(t, d)}$ is used. This is defined as the ratio of the maximum temperature rise encountered to the magnitude of the power steps producing the rise. The maximum temperature rise, and hence $R_{th(t, d)}$, depends on both the pulse duration t_p and the duty cycle d. The value of d is given by the equation $d = t_p/t_r$, where t_r is the periodic time. The value of $R_{th(t, d)}$ lies between the limits $R_{th(t,0)}$ and R_{th}. If the cooling period $t_r - t_p$ is greater than the temperature stabilisation time t_s, the junction always cools to its initial value between each pulse, and the temperature response is the same as that for a single step, that is $R_{th(t,d)} = R_{th(t,0)}$. If the heating time t_p is greater than the temperature stabilisation time, the steady-state junction temperature is reached, that is $R_{th(t,d)} = R_{th}$. For cooling periods shorter than t_s, the junction will not cool to its initial value and so the temperature excursions are superimposed on progressively higher values; this will continue until the temperature rise during the dissipating period is equal to the fall in temperature during the off period. The temperature reached at the end of any pulse after this quasi-equilibrium state is the maximum for this operation.

Methods of Examining Transient Thermal Resistance

Three methods of examining the transient thermal resistance are now described. The third method combines the two other procedures.

Theoretical Approach

If the heating curve of a semiconductor junction following a step function of power (Figure 32) could be regarded as a single exponential function, and if the initial and final temperatures were known, the specification of the thermal time constant would define the curve uniquely. With such a simple approximation the thermal response to power pulse trains could easily be calculated.

However, theoretical calculations considering one-dimensional heat flow indicate a more complex function (Ref. 91), in which the rate of increase in temperature is initially larger, such that the curve lies above that expected for exponential behaviour. This leads to a new definition of the time constant to be used in conjunction with this function: the time constant is that period required for the junction temperature to rise to 70% of its overall change as contrasted with approximately 63% for the true exponential.

Extension of the theoretical formula to include the temperature variation under pulsed excitation results in a family of loci relating the pulse duration, duty cycle, and transient thermal resistance, as in Figure 33. However, many simplifying assumptions are made in the theoretical transistor model, and in general theoretical curves tend to give an unduly large safety factor.

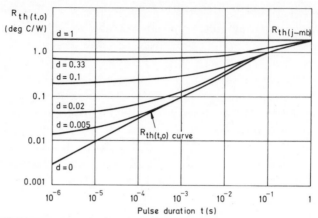

FIGURE 33.   Typical curves of transient R_{th} against pulse
duration.

Practical Approach

The direct practical method of determining the transient thermal resistance is to measure the actual junction temperature rise for specified values of pulse duration, duty cycle, and pulse power.   By making a large number of individual measurements, the pulse rating chart for a single transistor may be built up.   Such a chart, being based on one sample, cannot be used as a general rating for any particular type of device, as the spread for many devices must be taken into account.   In addition, life tests based on the pulse power calculated from the chart should indicate satisfactory serviceability of the device.

This method has been used to check the curves obtained otherwise, but it is a tedious method of producing a full chart.   It may be used to advantage, however, when an indication for a specific application is required.

Combined Approach

The basis of this approach is that the transient thermal resistance can be calculated from a knowledge of a single step in power, that is the curve corresponding to $d = 0$.   From the principle of the superposition of positive and negative power steps (Ref. 92), the following relation has been derived (see Ref. 90, Appendix 2):

$$R_{th(p\ d)} = dR_{th} + (1 - d)R_{th((t_r + t_p),0)} + R_{th(t_p,0)} - R_{th(t_r,0)} \qquad (13)$$

Where $R_{\text{th}(t_i, 0)}$ is the transient thermal resistance at the instant t_i, only the single-step response curve need be measured.

Use of Information

The information derived from the above calculations may be used in several ways. These are discussed below.

Calculation of Junction Temperature

The three common repetitive power waveforms shown in Figure 34 are considered.

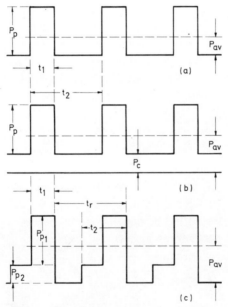

FIGURE 34. Common power waveforms.

Case 1. A power pulse train of magnitude P_p (Figure 34(a)) is dissipated in a device mounted on an infinite heatsink. The maximum junction temperature is then given by the equation

$$T_j = T_{\text{amb}} + P_p R_{\text{th}(t,d)} \qquad (14)$$

Under practical conditions, when the device is mounted on a heatsink of thermal resistance $R_{th(h)}$ with an interface thermal resistance of $R_{th(i)}$, the term $d(R_{th(h)} + R_{th(i)})P_p$ must be added. This is the rise in temperature of the mounting base due to the transmission of the average power to the ambient. Equation (14) then becomes

$$T_j = T_{amb} + P_p R_{th(t,d)} + d(R_{th(h)} + R_{th(i)}). \tag{15}$$

Case 2. The above power train is superposed on a steady dissipation level P (Figure 34(b)). The maximum junction temperature is then given by

$$T_j = T_{amb} + P_p R_{th(t,d)} + d(R_{th(h)} + R_{th(i)}) + P R_{th(j\ amb)}. \tag{16}$$

Case 3. The waveform in case 1 is superposed on a second repetitive rectangular waveform (Figure 34(c)) having the same repetition time but different pulse duration. It is important that individual pulses of each train finish simultaneously. Using the principle of superposition, the maximum junction temperature of a device mounted on an infinite heatsink is given by

$$T_j = T_{amb} + P_{p1} R_{th(t_1, d_1)} + P_{p2} R_{th(t_2, d_2)}, \tag{17}$$

where P_{p1}, t_1, and d_1 are the magnitude, pulse duration, and duty cycle respectively of the first pulse train, and P_{p2}, t_2, and d_2 the corresponding values of the superposed pulse train.

On a practical heatsink as in case 1 it becomes

$$T_j = T_{amb} + P_{p1} R_{th(t_1, d_1)} + P_{p2} R_{th(t_2, d_2)}$$
$$+ (d_1 P_{p1} + d_2 P_{p2})(R_{th(h)} + R_{th(i)}). \tag{18}$$

If, further, these trains are also superposed on a steady power P, the term $P R_{th(j-amb)}$ is simply added.

All the terms in the above expressions except $R_{th(t,d)}$ are defined uniquely. $R_{th(t,d)}$ may be found from the published data in two ways; if the relevant value of d is not one of the curves plotted, a rough interpolation may be made; alternatively, the following equation is provided which may be used to calculate $R_{th(t,d)}$ from the corresponding value of $R_{th(t,0)}$;

$$R_{th(t,d)} = d(R_{th(j-mb)} - R_{th(t,0)}) + R_{th(t,0)}. \tag{19}$$

This equation can be used to advantage as follows. In case 2 above, Equation (16) can be rewritten by substitution of $R_{th(t,d)}$;

$$T_j = T_{amb} + P_p \{d(R_{th(j-mb)} - R_{th(t-0)}) + R_{th(t-0)}\} + d(R_{th(h)} + R_{th(i)}) + \\ + PR_{th(j-amb)}$$

$$= T_{amb} + P_p(1 - d)R_{th(t,0)} + (P + dP_p)\,R_{th(j-amb)} \tag{20}$$

$$= T_{amb} + P(1 - d)\,R_{th(t,0)} + P_{av}R_{th(j-amb)}.$$

Similarly Equation (18) can be simplified to

$$T_j = T_{amb} + P_{p1}\,(1 - d_1)\,R_{th(t_1,0)} + P_{ps}\,(1 - d_2)\,R_{th(t_2,0)} + \tag{21}$$

$$+ P_{av}R_{th(j-amb)}.$$

In these two equations the pulse components are separated entirely from the steady-state thermal resistance, and average power; $R_{th(t,0)}$ is the quantity required from the chart, and separation of the components of $R_{th(j\_amb)}$ is not required.

In many practical cases the waveform is not a rectangle and the reader is referred to Ref. 90 for treatment of these conditions.

Practical Examples of Use of Rating Chart

Maximum Permissible Ambient Temperature

In this example a power supply of a maximum output rating of 30 V and 6 A is used to determine the maximum permissible ambient temperature. The power supply contains a bistable circuit, which for normal operation is in a particular state. When a short-circuit occurs across the output terminals, the large transient current that flows develops a voltage across a resistor. This voltage spike is sufficient to change the state of the bistable circuits, resulting in the application of a reverse-bias voltage across the emitter–base diode of the output transistors. Hence the output transistors are cut off and the output current falls to zero.   The bistable circuit, therefore, protects the output transistors and also the circuit which is supplied by the power unit.   On removal of the short circuit the bistable circuit can be reset and the power supply is again available for normal use.

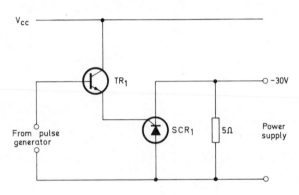

FIGURE 35.   Arrangement used to provide short circuit.

The power supply was set at 30 V and 6 A and then an effective short-circuit was applied across the output terminals using a thyristor, as shown in Figure 35.   The thyristor was triggered by single short pulses from a pulse generator, and the resulting waveforms, shown in Figure 36, of collector and emitter voltages and collector current of the six output transistors were obtained from photographs of the display from an oscilloscope.   The power pulse in each device was obtained by taking the collector–emitter voltage and collector current from the photographs at specific time intervals and is shown in Figure 37; the equivalent rectangle is also shown.

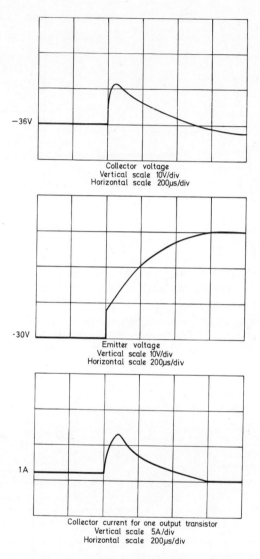

FIGURE 36.   Waveforms of current and voltage for the output transistors under short-circuit conditions.

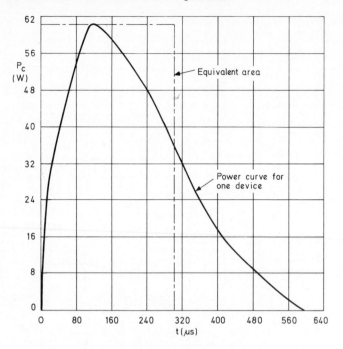

FIGURE 37.   Power dissipation in one output transistor under short-circuit conditions.

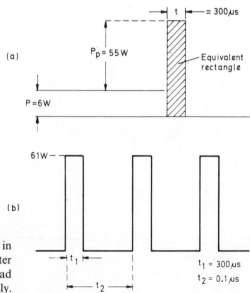

FIGURE 38. Pulse trains used in examples: (a) short-circuit load after long operation; (b) short-circuit load with supply being reset continuously.

Example 1. One of the conditions likely to occur is the short-circuiting of the output of a power supply. In the worst case this happens when the supply has been operating long enough to ensure that the output transistors have reached the maximum temperature indicated by the overall steady-state thermal resistance. The waveform, modified in such a way that the final pulse is rectangular, is shown in Figure 38(a).

The output transistors used in the supply show a transient thermal response similar to that plotted in Figure 39. From this it follows that the transient thermal resistance $R_{th(t,0)}$ of a single pulse of duration 300 μs is thus 0·13 degC/W, and that the steady-state thermal resistance $R_{th(j\ mb)}$ is 2 degC/W. The transistors are mounted on a heatsink for which $R_{th(h)}$ + $R_{th(i)}$ is 1·8 degC/W.

The temperature of the junction of one device is then given by

$$T_j = T_{amb} + PR_{th(j-amb)} + P_pR_{th(t,0)}$$
$$= T_{amb} + 6(2 + 1·8) + 55 \times 0·13$$
$$= (T_{amb} + 30) \ °C.$$

(Note that P_p is 55 W (not 61) since the principle of superposition of power pulses requires that each power component must finish at the time when T_j is measured. Thus 6 W of the short-circuit pulse is regarded as forming the final part of the steady-state component.)

Thus $T_{j(max)}$ is published as 90 °C; the power supply can operate under the above condition in a maximum ambient of 60 °C.

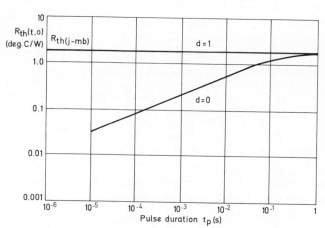

FIGURE 39.   Typical transient R_{th} curve assumed for transistor used.

Example 2. Another possible condition is for the short-circuit to be permanent, with the operator continually resetting the supply. It is to be assumed that the supply can be reset in 100 ms, which, from experience, is very optimistic. The resulting waveform, with each pulse modified to the equivalent rectangle as before, is shown in Figure 38(b). This train of pulses is dealt with as in case 1 (page 58) using Equation (15), that is

$$T_j = T_{amb} + P_p R_{th(t,d)} + d(R_{th(h)} + R_{th(i)}).$$

To find $R_{th(t,d)}$ Equation (19) is used, that is

$$R_{th(t,d)} = d(R_{th(j-mb)} - R_{th(t,0)}) + R_{th(t,0)}$$

where $d = t_1/t_2$.

By substitution, the following is obtained:

$$T_j = T_{amb} + P_p(1-d)R_{th(t,0)} + dP_p R_{th(j-mb)} \text{ (compare Equation (20))}$$

$$= T_{amb} + 61\left(1 - \frac{0\cdot0003}{0\cdot1}\right) \times 0\cdot13 + \frac{0\cdot003}{0\cdot1}\, 61(2 + 1\cdot8)$$

$$= (T_{amb} + 8\cdot6)\ ^\circ C.$$

Thus the power supply could operate in a maximum ambient temperature of 81·4 °C to observe a 90 °C limit for T_j.

In both of the above examples the transistor voltage and current ratings are observed, and it is also apparent that the junction temperature rating is observed for all normal ambient temperatures.

Maximum Permissible Pulse Power

This example shows the use of the transient thermal resistance chart to determine the maximum permissible pulse power for a device.

When a device has to dissipate pulse power as well as steady-state power, the relation between maximum permissible pulse power $P_{p(max)}$ and junction temperature is given by

$$P_{p(max)} = \frac{(T_{j(max)} - T_{(amb)}) - P_s R_{th(j-amb)}}{R_{th(t)} + d(R_{th(i)} + R_{th(h)})}, \tag{22}$$

where P_s is the steady-state dissipation of the device, $R_{th(t)}$ the effective transient thermal resistance of the device between junction and mounting base (it varies with the duty cycle of the pulse as shown by the curves in Figure 40), and d the duty cycle (the pulse duration t divided by the periodic time T).

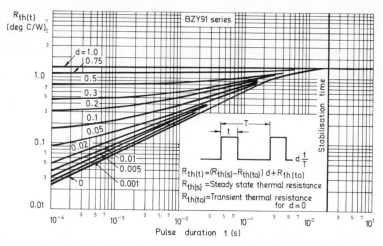

FIGURE 40.   Transient thermal resistance for various duty factors plotted against pulse duration.

As an example of how to determine the maximum pulse power, consider a voltage regulator diode type BZY91-C39, mounted on a heatsink in free air at 30 °C.   Suppose that the steady-state current is 500 mA, the duty cycle 0·1, and the pulse duration 1 ms.

The power dissipated by the voltage regulator diode is $V_Z I_Z$.   The heat generated within the diode, however, causes a change in the voltage V_Z across the diode (see   Effects of Temperature Changes , page 12).   Consequently, the dissipation at a constant current I_Z changes until the diode attains a steady temperature.   When the diode has reached a steady temperature, the voltage across it is given by

$$V_Z = \frac{V_{Zs} + (I_Z - I_{Zs})r_{Zs}}{1 - I_Z R_{th(j-amb)}},$$

where V_{Zs} is the voltage across the diode when the pulsed current through it is I_{Zs}, and S_Z the temperature coefficient of the diode.

If the heatsink consists of a single cooling fin, 100 cm^2 of 17 mm copper, $R_{th(h)}$ is 4 degC/W and

$$R_{th(j-amb)} = 1\,47 + 0\cdot2 + 4$$
$$= 5\cdot67 \text{ degC/W}.$$

Published data show that, when the pulsed current through the BZY91-C39 is 500 mA, the maximum voltage across it could be 41 V. Under the same condition, r_z is 1·4 Ω maximum and a typical value for S_z is 35 mV/degC. Hence,

$$V_Z = \frac{41 + (0\cdot5 - 0\cdot5)1.4}{1 - 0\cdot5 \times 5\cdot67 \times 35 \times 10^{-3}}$$

$$= 45\cdot5 \text{ V}$$

$$P_s = 0\cdot5 \times 45\,5$$

$$= 22\cdot75 \text{ W}.$$

Therefore, maximum permissible pulse power in this example is

$$P_{p(max)} = \frac{(175 - 30) - 22\cdot75 \times 5\cdot67}{0\,22 + 0\cdot1(0\cdot2 + 4\cdot0)}$$

$$= 25 \text{ W},$$

and the total permissible peak power is

$$P_s + P_{p(max)} = 47\cdot75 \text{ W}.$$

CHAPTER 3.   RECTIFICATION

Rectifications is the process of converting alternating current or voltage into direct current and voltage.   The wave shape of alternating voltage may be that of a sine wave, such as is common to 50 and 60 Hz mains supplies and motor generators; a square wave as found in the case of most d.c. inverters; or it may be any other shape as in the case of e.h.t. generators, ringing choke inverters, pulse width modulated switching circuits, etc.

In order to convert alternating current into direct current a device such as a rectifier diode is included in the circuit.   This is known as rectification.

It is the object of this chapter to show that the theory of rectification by means of vacuum tubes, which has been established for many years, equally applies to silicon rectifier diodes (Refs. 2, 93).   Apart from the more obvious advantages of considerable reduction in size and weight, silicon rectifier diodes have proved to be efficient and reliable and they require very little maintenance.   However, they have limited overload current capacity and are sensitive to overvoltage peaks of even a short duration.   It is therefore necessary to choose rectifier diodes carefully for any particular application, bearing in mind the operational as well as transient surges that are likely to occur (Refs. 94 to 106).

It is perfectly feasible to design high-power equipment using silicon rectifier diodes for powers up to a megawatt or more.   The practical approach to rectification in this book will be limited to the lower powers, although the treatment will also apply for the higher powers.

Throughout the first part of the chapter it has been assumed that the rectifier diodes are operated at low frequencies, that is below 400 Hz, which is the frequency limit normally quoted for published ratings of silicon rectifier diodes.

In certain applications, namely inverters and converters, it is necessary to operate rectifier diodes at high frequencies.   For high-frequency operation the average current that may be passed through the rectifier diode must be reduced from the low-frequency rating because of the heating effect caused by the minority carrier current flowing during the recovery time of the diode.

High-frequency operation will be discussed in a separate section of this chapter. For such application normally special rectifier diodes with short recovery times are used.

An assumption is made in this chapter that the rectifier diodes are ideal in the reverse direction. The leakage current will be considered in the discussion of series operation.

A comprehensive table of rectifier circuit characteristics will also be included.

Single-phase Circuits

Single-phase circuits are divided into five categories. These are (i) half-wave, (ii) full-wave centre-tap, (iii) full-wave bridge, (iv) voltage doubler, and (v) voltage multiplier circuits. The full-wave centre-tap circuit is also known as the two-phase half-wave circuit, but this form will not be used. For simplicity, however, the full-wave centre-tap circuit may be referred to as the full-wave circuit and the full-wave bridge circuit may be referred to as the bridge circuit. The voltage doubler circuit is treated separately, being a special case of the voltage multiplier circuits. The voltage tripler, voltage quadrupler, and higher-voltage circuits are treated under the common heading of the voltage multipliers.

Half-wave Circuit

The commonly used single-phase half-wave rectifier circuit with resistance load is shown in Figure 41. The output voltage waveform for the circuit is shown in Figure 42. The sine wave of the amplitude $E_{T(max)}$ shown in Figure 42(a) is the secondary output voltage of the transformer. This voltage is applied to the rectifier diode D as shown in the circuit diagram. The diode symbol indicates conventional current flow from anode to cathode. The diode conducts during the positive half-cycle and blocks during the negative half-cycle of the applied alternating voltage. Therefore the result is a discontinuous voltage and current in the load.

To provide continuous load current a capacitor input filter is used as shown in Figure 43.

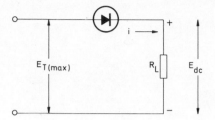

FIGURE 41. Half-wave rectifier circuit.

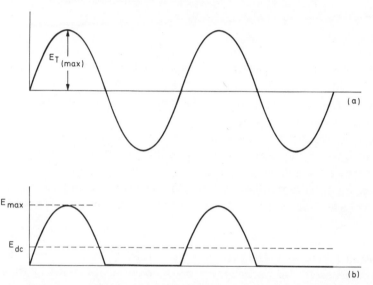

FIGURE 42. Waveforms for single-phase half-wave circuit: (a) sine-wave input; (b) rectifier output.

The capacitor charges to the crest value of the applied voltage on the first positive half-cycle. When the applied voltage falls below the crest value, the voltage across the capacitor is higher than the applied voltage so that the rectifier diode is reverse-biased. From that time and during the negative half-cycle, the capacitor discharges into the load. As the next positive half-cycle is applied, the diode becomes forward-biased again when the input voltage exceeds the capacitor voltage. The rectifier diode conducts and charges the capacitor to the crest applied voltage. The diode stops conducting again and the cycle is repeated. This continues for as long as there is the sine-wave voltage at the input.

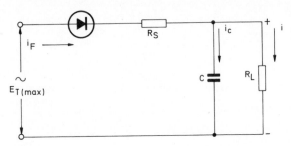

FIGURE 43. Half-wave circuit with capacitor input filter.

The capacitor C is initially uncharged and the load on the rectifier is therefore effectively a short-circuit. The current through the rectifier under this condition is limited only by the source resistance of the secondary winding of the transformer.

A series resistor R_s is therefore included in the circuit to limit the peak current through the rectifier diode on initial switch-on. The value of R_s may include the source resistance for calculation purposes.

The total resistance, however, must not be made too large, as this results in a voltage drop across it which leads to a loss in efficiency and poor regulation.

The current waveforms after a steady state has been established are shown in Figure 44. The current waveform is idealised as it does not rise instantaneously. In practice there is a time constant formed by the capacitor C and the sum of the total circuit resistance which includes source resistance, the resistance R_s, and the resistance of the diode itself.

The purpose of the capacitor is to store up the energy during the conduction period of the rectifier diode and to supply the load current when the diode is not conducting. The diode current is equal to the sum of the load current and the capacitor current. The load current is then equal to the capacitor current during the non-conducting period.

The circuit suffers from a number of disadvantages which are high peaks of rectifier current and low ripple frequency requiring a large value of filter capacitor to keep the amplitude of the ripple voltage down. The ripple frequency, in fact, is equal to the frequency of the applied voltage.

Further disadvantages come about when a transformer is used to supply the power at a required voltage level to the rectifier circuit. The secondary of the transformer carries undirectional current and affects the flux in the

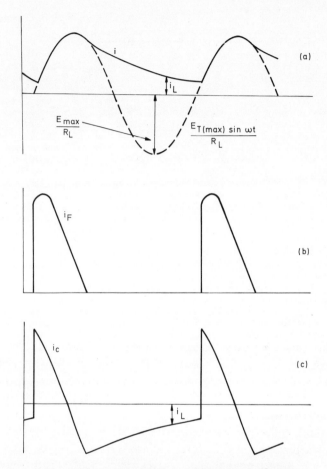

FIGURE 44. Waveforms for single-phase half-wave circuit, after establishment of steady state: (a) load current; (b) rectifier current; (c) capacitor current.

core. This in turn increases the magnetising current and core losses, at the same time introducing harmonics in the secondary voltage. The utility factor of the transformer is also low.

The half-wave circuit is therefore most commonly used direct from the mains.

The regulation of the circuit is poor and the conversion efficiency is low.

The single-phase half-wave circuit is only used with a capacitor input filter. A choke input filter would require a large value of inductance to cause the current to flow through the cycle and therefore is ruled out.

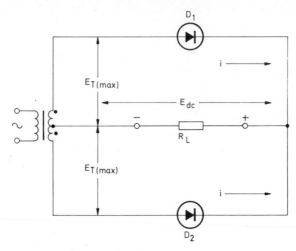

FIGURE 45.   Full-wave centre-tap circuit.

Full-wave Circuit

The full-wave centre-tap circuit is shown in Figure 45. The rectifier diodes are connected so that the diode D_1 conducts when the end of the winding marked with a dot goes positive. At the same time the end of the winding to which the diode D_2 is connected goes negative so that the diode D_2 does not conduct.

During the following half-cycle the polarities of the applied voltage reverse. The diode D_1 is then reverse-biased owing to the negative voltage applied to it and does not conduct, whereas the diode D_2 is forward-biased owing to the positive voltage applied to it and is conducting as shown in Figure 46. The letters A and B (used for D_1 and D_2 respectively) in Figure 46(b) indicate which diode is conducting at that time. The diodes conduct alternately and therefore current flows through each of the transformer secondaries alternately.

In the full-wave circuit the rectifier diodes must withstand a crest working voltage which is equal to the peak value of the applied voltage across both halves of the transformer secondary, i.e. $2E_{T(max)}$.

The current through the load is unidirectional. To avoid discontinuous, or pulsating, voltage and current at the load, a capacitor C is connected across the load to smooth the output in a similar manner to that described for the half-wave circuit.

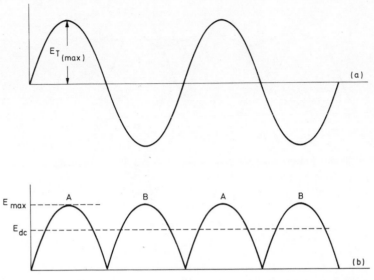

FIGURE 46.   Waveforms for full-wave circuit: (a) sine-wave input;
(b) rectifier output.

The full-wave circuit with a capacitor input filter is shown in Figure 47. The capacitor is charged during part of each half-cycle so that it has to maintain the load current for a shorter period compared with the half-wave circuit. The ripple frequency is twice that of the applied voltage which means that a capacitor of equal value is twice as effective. The capacitor voltage changes by a smaller amount and therefore the available d.c. output voltage is greater than that for the half-wave circuit and the amplitude of the ripple voltage is smaller.

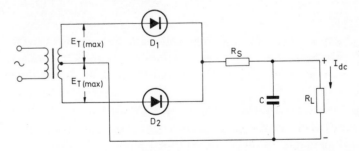

FIGURE 47.   Single-phase full-wave centre-tap circuit (also
known as two-phase half-wave).

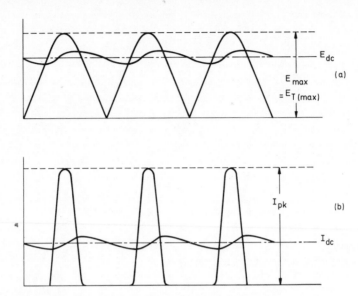

FIGURE 48. Waveforms for single-phase full-wave circuits:
(a) voltage; (b) current.

The output waveforms are shown in Figure 48.

The full-wave circuit, unlike the half-wave circuit, can never be used without a transformer on single phase.

The full-wave circuit is used for low-power low-voltage applications where low ripple or high efficiency is desired.

Bridge Circuit

The bridge circuit shown in Figure 49 is another version of the full-wave circuit. In the bridge circuit, rectifier diodes D_2 and D_3 conduct during the positive half-cycle, and diodes D_1 and D_4 conduct during the negative half-cycle.

The performance of the bridge circuit is the same as that of the full-wave centre-tap circuit, except that rectifier crest working voltage, in a bridge circuit is half that of the full-wave centre-tap circuit for the same d.c. voltage.

The voltage and current waveforms for the bridge circuit are the same as that for the full-wave circuit shown in Figure 48.

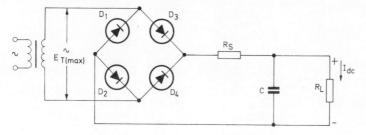

FIGURE 49.   Single-phase full-wave bridge circuit.

The bridge circuit is more widely used than the full-wave circuit because of the higher power capability.   It is also used wherever the desired output voltage is approximately equal to the applied voltage.

The major advantage of the bridge circuit is that only one secondary winding is required, reducing wastage in transformer insulation and winding space, resulting in higher secondary utility factor.

Voltage Doubler Circuits

Voltage doubler circuits are in fact the lowest order of the voltage multipliers discussed later.   As they are better known and are more frequently used, they are treated separately.

There are three types of voltage doubler circuits: (i) symmetrical voltage doubler, (ii) common-terminal voltage doubler, otherwise known as the diode pump, and (iii) bridge voltage doubler.

The three circuits are similar in that they produce approximately the same output voltage from an equal transformer winding.   They differ mainly in the ratings of the rectifier diodes and the capacitors, as well as in the ripple frequency of the output voltage.

Symmetrical Voltage Doubler

The symmetrical voltage doubler circuit shown in Figure 50 is a combination of two half-wave rectifier circuits and is virtually a reversal of the full-wave centre-tap circuit.

Instead of the two rectifiers feeding a common smoothing capacitor from separate transformer windings, in the symmetrical voltage doubler

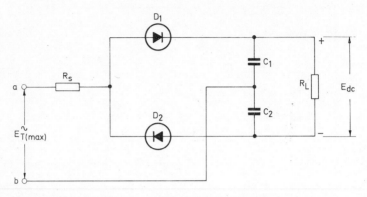

FIGURE 50.   Symmetrical voltage doubler circuit.

the two rectifier diodes feed separate smoothing capacitors connected in series, but are supplied from the same input.

The operation of the circuit is as follows.   When the top of the input is positive, current flows through R_s and diode D_1 to charge the capacitor C_1. The other terminal of the capacitor C_1 is connected to the bottom end of the input.

When the bottom end of the winding is positive, the capacitor C_2 is charged by the current which now flows through R_s and diode D_2.

Each capacitor is charged to the peak value of the applied voltage.   The output voltage therefore is twice the peak applied voltage.   This value of the output voltage, however, can only be achieved if the load is disconnected. Since the load is permanently connected across the two capacitors, the capacitors continually discharge in alternate half-cycles.   If the capacitor values are large enough, they act as smoothing elements and the voltage fall is small.

The final voltage to develop across the capacitors depends upon their values and the load.   The ripple frequency in the output voltage is twice that of the applied voltage.   The output voltage waveform for the symmetrical voltage doubler circuit is shown in Figure 51.

The rectifier diodes must be rated at twice the peak applied voltage and the capacitors must be rated at the peak applied voltage.

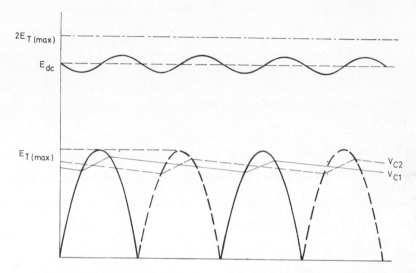

FIGURE 51.   Output voltage waveform for symmetrical voltage doubler.

Common-terminal Voltage Doubler

The common-terminal voltage doubler is shown in Figure 52. The circuit is sometimes referred to as a diode pump, since the voltage across C_2 is pumped up to approximately twice the peak input voltage during each cycle.

The operation of the circuit is best explained starting with the negative half-cycle. During this time the capacitor C_1 is charged to the peak input

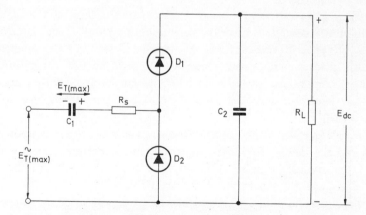

FIGURE 52.   Common-terminal voltage doubler circuit.

voltage $E_{T(max)}$ through the rectifier diode D_2. When the next positive half-cycle follows, the voltage across C_1 is in series with the applied voltage. D_1 now conducts and charges the capacitor C_2 to $2E_{T(max)}$. C_1 in the mean-time discharges somewhat owing to the load R_L and to charge the capacitor C_2, but charges again to $E_{T(max)}$ during the next negative half-cycle. The cycle is then repeated. The capacitor C_2 in turn supplies the load current during the negative half-cycle so that the voltage across it, E_{dc}, does not in fact reach the value of $2E_{T(max)}$ as shown in Figure 53.

The ripple frequency of the common-terminal voltage doubler is the same as that of the applied voltage. The regulation of the circuit is therefore worse than that of the symmetrical circuit and higher values of the capacitors are required. Capacitor C_1 must be rated at the peak applied voltage and must be capable of carrying the r.m.s. load current. Capacitor C_2 must be rated at twice the peak applied voltage. The rating of the rectifier diode is the same as that for the symmetrical circuit, that is they must withstand twice the peak applied voltage.

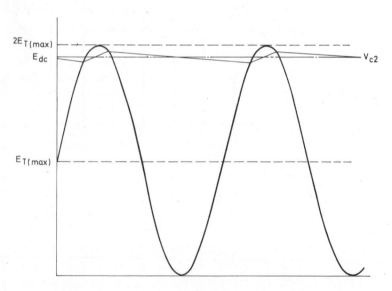

FIGURE 53.   Output voltage waveform for common-terminal voltage doubler.

Bridge Voltage Doubler Circuit

The bridge voltage doubler circuit shown in Figure 54 is a combination of the full-wave bridge circuit with a symmetrical voltage doubler circuit. This results in a more stable circuit with better regulation than the symmetrical circuit. The diodes D_3 and D_4 help in supplying load current during alternate half-cycles at the instances when capacitors C_1 and C_2 in the symmetrical voltage doubler circuit are being discharged. The output voltage is therefore closer to $2E_{T(max)}$ and the circuit is capable of higher load current.

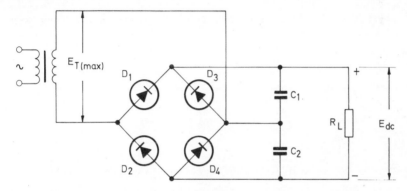

FIGURE 54. Single-phase full-wave bridge voltage doubler circuit.

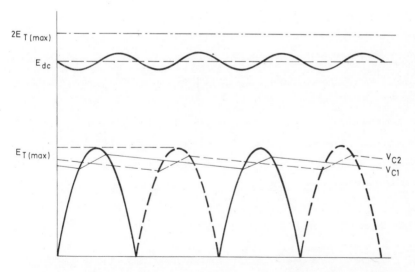

FIGURE 55. Output voltage waveform for bridge voltage doubler circuit.

The voltage waveforms shown in Figure 55 are the same as in the symmetrical voltage doubler and the ripple frequency is also twice that of the applied voltage.

The rating of the rectifier diodes, despite the bridge connection, is that of twice the peak applied voltage. This is due to the fact that the junction of the diodes D_3 and D_4 are strapped to the junction of the capacitors C_1 and C_2 and thus bypass the a.c. signal across the rectifiers D_1 and D_2.

Voltage Multipliers

The voltage multipliers include such circuits as voltage triplers, voltage quadruplers, and the higher-order circuits. There are several ways in which these circuits can be arranged.

Voltage doubler circuits have already been discussed. The circuits that follow in this section lead to the Cockcroft–Walton series multiplier.

Voltage Tripler

In the case of voltage triplers, there are three possible arrangements. A circuit shown in Figure 56 is a combination of a common-terminal voltage doubler with a half-wave rectifier circuit. The capacitors C_1 and C_2 and diodes D_1 and D_2 form a voltage doubler as shown in the previous section.

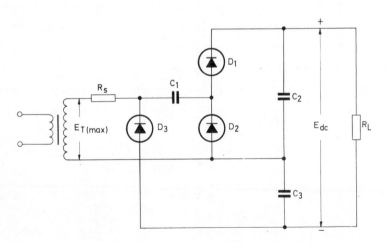

FIGURE 56.   Voltage tripler circuit.

Adding a capacitor C_3 and a diode D_3 makes the circuit into a voltage tripler.

The other two voltage tripler circuits shown in Figures 57 and 58 are very similar except that they are both common-terminal circuits in which the load and one side of the transformer secondary winding are joined together and therefore can be connected to a common earthing point. The only difference between the two common-terminal arrangements is that in Figure 57 the load is connected across C_2 and C_3 which are in series and therefore require lower individual voltage ratings, whereas in Figure 58 the output voltage appears across C_3 which has to be rated at three times the peak input voltage. This is a serious disadvantage as the price of the capacitor depends on its voltage rating, so that for very high voltages the configuration of Figure 58 would be more expensive than that of Figure 57. The circuit of Figure 58 has no real advantages over the circuit of Figure 57 and therefore is not used in practice.

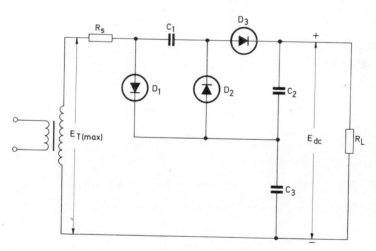

FIGURE 57. Common-terminal voltage tripler.

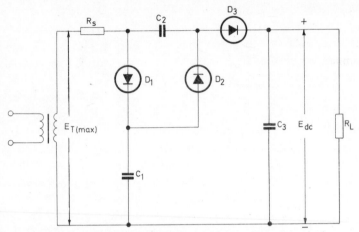

FIGURE 58. Common-terminal voltage tripler with single-capacitor output.

Voltage Quadrupler Circuits

Using four capacitors and four diodes a number of different voltage quadrupler circuits are obtained. Some of these are shown in Figures 59, 60, and 61. The circuits are basically combinations of two common-terminal voltage doubler circuits. The two input capacitors C_1 and C_3 are connected to give a symmetrical voltage quadrupler circuit shown in Figure 59. Two common-terminal circuits are shown in Figures 60 and 61.

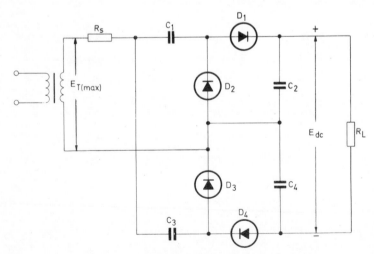

FIGURE 59. Symmetrical voltage quadrupler circuit.

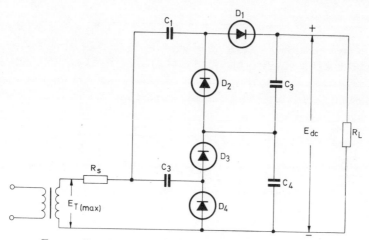

FIGURE 60.   Common-terminal parallel quadrupler circuit.

The symmetrical circuit is best as far as the capacitor voltage rating is concerned, as all capacitors are rated at twice the peak input voltage.

In the common-terminal circuit shown in Figure 60 the capacitor C_1 is rated at three times the peak input voltage and the capacitor C_3, is rated at the peak applied voltage.   The capacitors C_1 and C_3 are also sharing the r.m.s. load current.   The remaining two capacitors C_2 and C_4 are rated at twice the peak applied voltage.

In Figure 61 the voltage rating of the capacitors is more evenly distributed. C_3 is rated at the peak applied voltage but it has to cope with full r.m.s. load current.   The remaining capacitors are rated at twice the peak applied voltage.

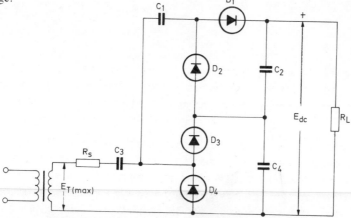

FIGURE 61.   Common-terminal series quadrupler circuit.

Voltage Multipliers

The above-described voltage doubler, tripler, and quadrupler circuits can be extended to multiply the rectified half-wave n times. The circuit is commonly known as Cockroft–Walton. A common-terminal series voltage multiplier with n half-wave stages is shown in Figure 62. This configuration is for odd numbers of n stages.

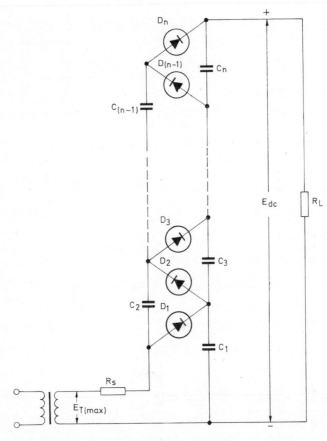

FIGURE 62. Common-terminal series voltage multiplier with n stages (where n is odd).

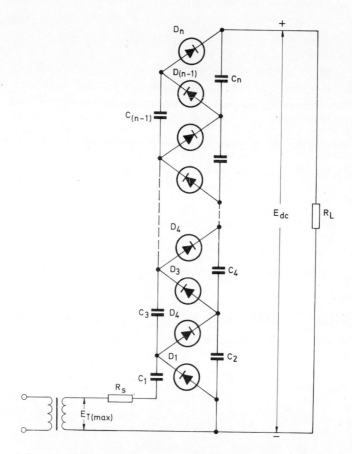

FIGURE 63.   Common-terminal series voltage multiplier
with $\frac{1}{2}n$ doubling stages (where n is even).

If the positions of C_1 and D_1 are interchanged with respect to the secondary winding of the transformer, a circuit of a common-terminal series voltage multiplier with n even number of half-wave stages is obtained.   The circuit as shown in Figure 63 is virtually a connection of $\frac{1}{2}n$ voltage doubling stages.

Whether odd or even multiples of the peak applied voltage appear across the load depends on which side of the ladder circuit connected to the secondary of the transformer is made to be common with the load.   This is shown in Figures 62 and 63.

Another higher-order voltage multiplier circuit is shown in Figure 64. The circuit is based on the common-terminal voltage doubler circuit in conjunction with symmetrical and common-terminal voltage quadrupler circuits.   The output voltage is six times the peak applied voltage.

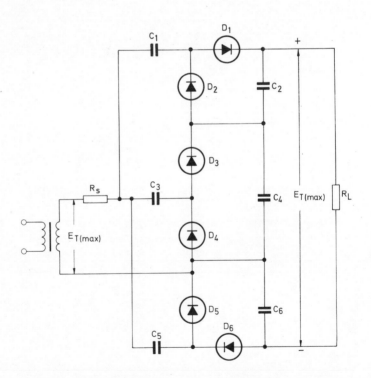

FIGURE 64.   Symmetrical voltage multiplier ($E_{dc} = 6E_T$).

The output voltages as quoted for the above circuits apply only for no load conditions.   As soon as a load is connected, the actual voltage will depend on the component values and the regulation of the circuit used.

The idealised rectifier circuit performances are given in Table 1 on pages 88-9 (see also Refs. 2, 106).

Design Procedure for Circuit with Capacitor Input Filter

The graphical analyses carried out by Schade (Ref. 93) apply to silicon rectifiers with little or no modification.   The design charts have been adapted, with permission, from the original article for the design procedure given below.

Unlike the case of thermionic valve rectifiers, the forward voltage drop across the silicon rectifier diodes is small.   The voltage drop varies by a small amount with forward current, and the increase in the forward voltage may be ignored without any loss of accuracy because in most circuits this represents a very low percentage of the output voltage.   Therefore, for the purpose of calculation, the forward voltage drop is considered to be that which occurs across the rectifier when the maximum required average current is flowing through the rectifier.

Other factors to be considered when designing any rectifier circuit concern the published data.   The four main characteristics not to be exceeded in circuits with a capacitor input filter are (i) maximum crest working voltage rating of the rectifier, (ii) initial switch-on peak current through the rectifier, (iii) repetitive peak current through the rectifier, and (iv) ripple current through the capacitor.

In the graphical design solution which follows, the source resistance R_s includes the transformer winding resistance, the rectifier resistance, and the series resistance, if required, to limit the initial peak rectifier current. The transformer leakage reactance has not been taken into account in the design procedure.   It tends to reduce the peak rectifier current and therefore assists in limiting the peak current even if additional series resistance is not included.

Figures 65, 66, and 67 give the conversion ratio $E_{dc}/E_{T(max)}$ as a function of $\omega R_L C$ for half-wave, full-wave and voltage doubler circuits respectively. The conversion ratio depends on the value of $R_s/R_L \%$. For reliable operation the value of $\omega R_L C$ should be selected to allow operation on the flat portion of the curves.

TABLE 1. Idealised rectifier circuit performances

	Single – phase			Three – phase			
Type of rectifier circuit	Half – wave	Centre – tap full – wave	Full – wave bridge	Half – wave	Full – wave bridge	Centre – tap	Double star with interphase transformer
Secondary input voltage per phase	$E_{max} = E_{T(max)}$ $E_{rms} = 0.707 E_{T(rms)}$	$E_{max} = E_{T(max)}$ $E_{rms} = E_{T(rms)}$	$E_{max} = E_{T(max)}$ $E_{rms} = E_{T(rms)}$	$E_{max} = E_{T(max)}$ $E_{rms} = 1.2 E_{T(rms)}$	$E_{max} = (3)^{1/2} E_{T(max)}$ $E_{rms} = 2.34 E_{T(rms)}$	$E_{max} = E_{T(max)}$ $E_{rms} = 1.35 E_{T(rms)}$	$E_{max} = 0.866 E_{T(max)}$ $E_{rms} = 1.17 E_{T(rms)}$
Output voltage across a–b							
Number of output voltage pulses per cycle	1	2	2	3	6	6	6
Voltage relationships							
Crest working voltage in terms of E_{dc}	$3.14 E_{dc}$	$3.14 E_{dc}$	$1.57 E_{dc}$	$2.09 E_{dc}$	$1.05 E_{dc}$	$2.09 E_{dc}$	$2.42 E_{dc}$
Crest working voltage in terms of $E_{T(rms)}$	$1.41 E_{T(rms)}$	$2.82 E_{T(rms)}$	$1.41 E_{T(rms)}$	$2.45 E_{T(rms)}$	$2.45 E_{T(rms)}$	$2.83 E_{T(rms)}$	$2.83 E_{T(rms)}$
E_{dc} in terms of r.m.s input voltage per phase $E_{T(rms)}$	$0.45 E_{T(rms)}$	$0.90 E_{T(rms)}$	$0.90 E_{T(rms)}$	$1.17 E_{T(rms)}$	$2.34 E_{T(rms)}$†	$1.35 E_{T(rms)}$	$1.17 E_{T(rms)}$
E_{dc} in terms of r.m.s output voltage E_{rms}	$0.636 E_{rms}$	$0.90 E_{rms}$	$0.90 E_{rms}$	$0.98 E_{rms}$	E_{rms}	E_{rms}	E_{rms}
E_{dc} in terms of peak output voltage E_{max}	$0.318 E_{max}$	$0.636 E_{max}$	$0.636 E_{max}$	$0.826 E_{max}$	$0.955 E_{max}$	$0.955 E_{max}$	$0.955 E_{max}$
Input voltage $E_{T(rms)}$ in terms of E_{dc}	$2.22 E_{dc}$	$1.11 E_{dc}$	$1.11 E_{dc}$	$0.855 E_{dc}$	$0.428 E_{dc}$†	$0.74 E_{dc}$	$0.855 E_{dc}$
r.m.s output voltage E_{rms} in terms of E_{dc}	$1.57 E_{dc}$	$1.11 E_{dc}$	$1.11 E_{dc}$	$1.02 E_{dc}$	$1.00 E_{dc}$	$1.00 E_{dc}$	$1.00 E_{dc}$
Peak output voltage E_{max} in terms of E_{dc}	$3.14 E_{dc}$	$1.57 E_{dc}$	$1.57 E_{dc}$	$1.21 E_{dc}$	$1.05 E_{dc}$	$1.05 E_{dc}$	$1.05 E_{dc}$

88

TABLE 1. (Continued)

Fundamental ripple frequency f %ripple = $\dfrac{\text{r.m.s fundamental ripple voltage}\times100}{E_{dc}}$		f 111	$2f$ 47.2	$2f$ 47.2	$3f$ 17.7	$6f$ 4.0	$6f$ 4.0	$6f$ 4.0
Output current								
Average current per rectifier leg $I_{F(av)}$		I_{dc}	$0.5I_{dc}$	$0.5I_{dc}$	$0.33I_{dc}$	$0.33I_{dc}$	$0.167I_{dc}$	$0.167I_{dc}$
$I_{r.m.s}$ per rectifier leg	R	$1.57I_{dc}$	$0.785I_{dc}$	$0.785I_{dc}$	$0.588I_{dc}$	$0.577I_{dc}$	$0.408I_{dc}$	$0.293I_{dc}$
	L		$0.707I_{dc}$	$0.707I_{dc}$	$0.577I_{dc}$	$0.577I_{dc}$	$0.408I_{dc}$	$0.289I_{dc}$
I_{pk} per rectifier leg	R	$3.14I_{dc}$	$1.57I_{dc}$	$1.57I_{dc}$	$1.21I_{dc}$	$1.05I_{dc}$	$1.05I_{dc}$	$0.605I_{dc}$
	L		I_{dc}	I_{dc}	I_{dc}	I_{dc}	I_{dc}	$0.5I_{dc}$
Transformer ratings								
Secondary r.m.s voltage per transformer leg $E_{T(rms)}$		$2.22E_{dc}$	$1.11E_{dc}$ (to centre-tap)	$1.11E_{dc}$	$0.855E_{dc}$ (to neutral)	$0.428E_{dc}$† (to neutral)	$0.74E_{dc}$ (to neutral)	$0.855E_{dc}$ (to star point)
Secondary r.m.s current per transformer leg $I_{T(rms)}$	R	$1.57I_{dc}$	$0.785I_{dc}$	$0.785I_{dc}$	$0.588I_{dc}$	$0.816I_{dc}$†	$0.408I_{dc}$	$0.293I_{dc}$
	L		$0.707I_{dc}$	$0.707I_{dc}$	$0.577I_{dc}$	$0.816I_{dc}$†	$0.408I_{dc}$	$0.289I_{dc}$
Secondary volt-ampere VA_s	R	$3.48E_{dc}\times I_{dc}$	$1.74E_{dc}\times I_{dc}$	$1.23E_{dc}\times I_{dc}$	$1.50E_{dc}\times I_{dc}$	$1.05E_{dc}\times I_{dc}$	$1.8E_{dc}\times I_{dc}$	$1.50E_{dc}\times I_{dc}$
	L		$1.57E_{dc}\times I_{dc}$	$1.11E_{dc}\times I_{dc}$	$1.48E_{dc}\times I_{dc}$	$1.05E_{dc}\times I_{dc}$	$1.8E_{dc}\times I_{dc}$	$1.48E_{dc}\times I_{dc}$
Secondary utility factor U_s	R	0.287	0.574	0.813	0.666	0.95	0.552	0.666
	L		0.636	0.90	0.675	0.95	0.552	0.675
Primary voltage per transformer leg (transformer ratio 1:1)		$2.22E_{dc}$	$1.11E_{dc}$	$1.11E_{dc}$	$0.855E_{dc}$	$0.428E_{dc}$	$0.74E_{dc}$	$0.855E_{dc}$
Primary voltage per transformer leg (transformer ratio 1:1)	R	$1.57I_{dc}$	$1.11I_{dc}$	$1.11I_{dc}$	$0.588I_{dc}$	$0.816I_{dc}$	$0.577I_{dc}$	$0.408I_{dc}$
	L	I_{dc}	I_{dc}	I_{dc}	$0.471I_{dc}$	$0.816I_{dc}$	$0.577I_{dc}$	$0.408I_{dc}$
Primary volt-ampere VA_p	R	$3.48E_{dc}\times I_{dc}$	$1.23E_{dc}\times I_{dc}$	$1.23E_{dc}\times I_{dc}$	$1.50E_{dc}\times I_{dc}$	$1.05E_{dc}\times I_{dc}$	$1.28E_{dc}\times I_{dc}$	$1.05E_{dc}\times I_{dc}$
	L		$1.11E_{dc}\times I_{dc}$	$1.11E_{dc}\times I_{dc}$	$1.21E_{dc}\times I_{dc}$	$1.05E_{dc}\times I_{dc}$	$1.28E_{dc}\times I_{dc}$	$1.05E_{dc}\times I_{dc}$
Primary utility factor U_p	R	0.287	0.813	0.813	0.666	0.95	0.78	0.95
	L		0.90	0.90	0.827	0.95	0.78	0.95

† For three-phase full-wave bridge a delta-connected secondary can also be used. Then r.m.s. voltage per transformer leg $= 0.74E_{dc}$ and r.m.s. current per transformer leg $= 0.472I_{dc}$.

R, resistive load; L, inductive load; f, supply frequency (Hz).

In the calculation of the above circuit performances, the rectifier forward voltage drop and the transformer impedance have been ignored.

The primary volt–ampere rating of the transformer does not take primary magnetising current into account.

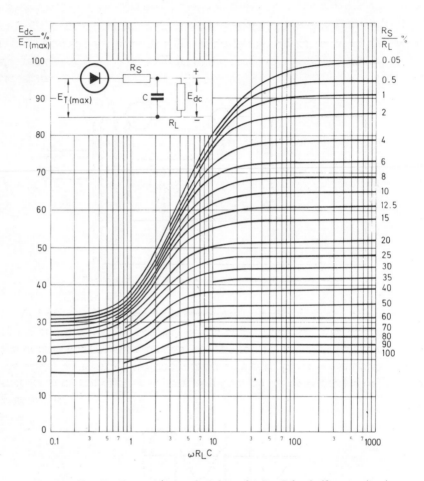

FIGURE 65. $E_{dc}\, E_{T(max)}$ % as a function of $\omega R_L\, C$ for half-wave circuits. C in farads; R_L in ohms; $\omega = 2\pi f$.

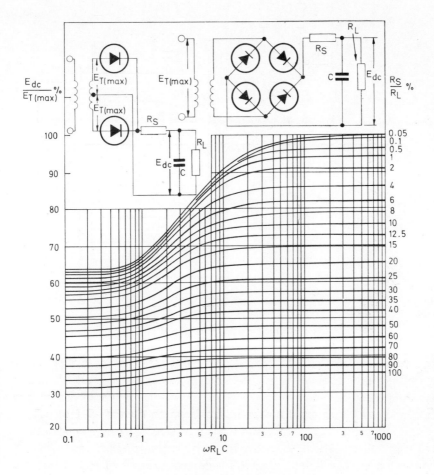

FIGURE 66. $E_{dc} E_{T(max)}$ % as a function of $\omega R_L C$ for full-wave circuits. C in farads; R_L in ohms; $\omega = 2\pi f$.

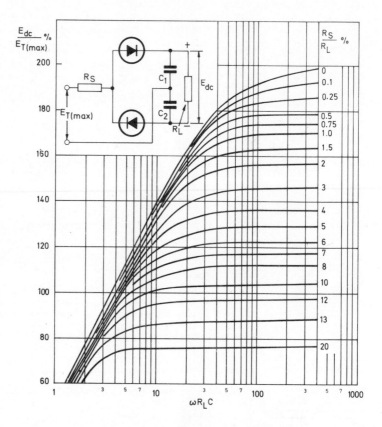

FIGURE 67. $E_{dc} E_{T(max)}$ % as a function of $\omega R_L C$ for voltage doubler circuits. C in farads; R_L in ohms; $\omega = 2\pi f$.

Figure 68 gives information on the minimum value of $\omega R_L C$ that must be used to reduce the percentage ripple to a desirable figure. Figures 69 and 70 give, respectively, the ratio of r.m.s. rectifier current to average current per rectifier and the ratio of peak repetitive rectifier current to average current per rectifier, plotted as functions of $n\omega R_L C$ These ratios are dependent on the value of R_s/nR_L%.

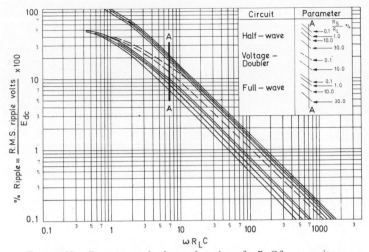

FIGURE 68. Percentage ripple as a function of $\omega R_L C$ for capacitor input filter. C in farads; R_L in ohms; $\omega = 2\pi f$ (f is the line frequency).

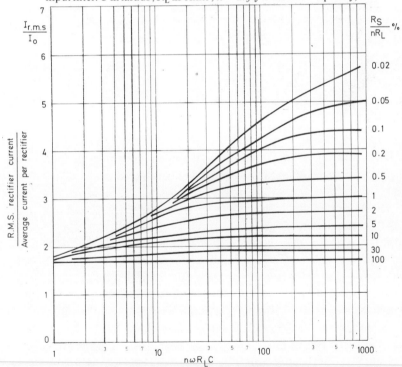

FIGURE 69. The ratio of r.m.s. rectifier current to average current per rectifier, plotted against $n\omega R_L C$. C in farads; R_L in ohms; $n = 1$ for half-wave, $n = 2$ for full-wave, $n = 0.5$ for voltage doubler.

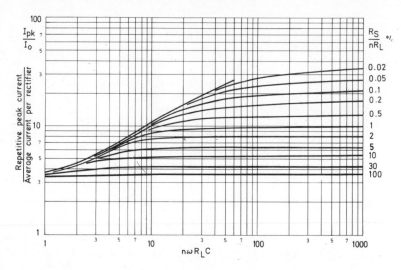

FIGURE 70. The ratio of repetitive peak current to average current per rectifier, plotted against $n\omega R_L C$. C in farads; R_L in ohms; $\omega = 2\pi f$ (f is the line frequency); $n = 1$ for half-wave, $n = 2$ for full-wave, $n = 0.5$ for voltage doubler.

Design Procedure

The following design procedure is recommended in the design of single-phase silicon rectifier circuits with capacitor input filter.

(1) Determine the value of R_L.

(2) Assume a value of R_s (usually between 1 and 10% of R_L).

(3) Calculate $R_s/R_L\%$.

(4) From the percentage ripple graph against $\omega R_L C$ (Figure 68), determine the value of $\omega R_L C$ required to reduce the ripple to a desired value for $R_s/R_L\%$ determined in (3). Calculate the value of C required.

(5) From the $(E_{dc}/E_{T(max)}\%, \omega R_L C)$ curves for the appropriate circuit (Figures 65, 66, or 67) determine the conversion ratio for the value of $\omega R_L C$ determined in (4) and $R_s/R_L\%$ determined in (3).

(6) Determine the $E_{T(max)}$ and $E_{T(rms)}$ that must be applied to the circuit using information derived in (5).

(7) Determine the crest working voltage that the rectifiers must withstand.

(8)   Determine the r.m.s. current per rectifier from Figure 69.

(9)   Decide on the rectifiers to be used.

(10)   Check the peak repetitive current per rectifier from Figure 70.

(11)   Check the initial switch-on current I_{on} given by $E_{T(max)}/R_s$. If the value obtained exceeds that specified for the rectifier, then R_s must be increased and the design procedure repeated.

(12)   Design the transformer and adjust the value of R_s accordingly, taking into account the transformer resistance and the forward resistance of the rectifier at the average current.

(13)   Check the r.m.s. ripple current through the capacitor.

(14)   Design the RC damping circuit as recommended in the published data of the rectifier.

(15)   Determine the size of heatsink to be used to allow operation a the desired ambient temperature (from published data).

When considering the rectifier ratings, it is also necessary to take into account the fluctuation in the alternating voltage and the input voltage distortion due to harmonics, as well as transients which are likely to occur on mains supplies.

In order to suppress the transients a series RC damping circuit may be used.   The values of R and C are calculated according to the information given in the rectifier data.

An important factor in a rectifier circuit design is the ripple current that flows through the reservoir capacitor.   The capacitor used in the circuit must be suitably rated to handle the ripple current.   The total r.m.s. current flowing through the reservoir capacitor $I_{c(rms)}$ can be calculated from the r.m.s. current flowing through each rectifier I_{rms} and the d.c. output current I_{dc}.

For single-phase half-wave and voltage doubler circuits

$$I_{c(rms)} = (I_{rms}^2 - I_{dc}^2)^{\frac{1}{2}}.$$ (23)

In the full-wave rectifier circuits, half the total r.m.s. current flows through each rectifier; therefore

$$I_{c(rms)} = (2I_{rms}^2 - I_{dc}^2)^{\frac{1}{2}}$$ (24)

This clearly indicates that the rectifier circuits using capacitor input filters are limited in their current-handling capacity.

Design Examples

The design of the four types of capacitor input filter rectifier circuit is illustrated in Table 2, the recommended procedure being used.

TABLE 2

Design examples for single-phase circuits with capacitor input filter

Requirement	Type of rectifier circuit			
	Half-wave	*Full-wave bridge*	*Centre-tap full-wave*	*Voltage doubler*
E_{dc} (V)	150	300	120	600
I_{dc}(A)	1·5	2·0	2·0	1·0
V_R (% ripple)	$\leqslant 1$	$\leqslant 1$	$\leqslant 1·3$	$\leqslant 1$
f (Hz)	50	50	50	50
Solution				
(1) Load resistance $R_L = E_{dc}/I_{dc}$ (Ω)	100	150	60	600
(2) Let source resistance R_s (Ω) be	6	9	3·6	12
(3) R_S/R_L (%)	6	6	6	2
(4) Value of $\omega R_L C$ from Figure 68	150	66	50	≈ 150
$C = \dfrac{(\omega R_L C)}{2\pi 50 R_L} (\mu F)$	4870	1400	2660	795
Practical value of C (μF)	5000	1800	3000	1000
New value of $\omega R_L C$	157	85	56·5	188

TABLE 2 (*continued*)

Requirement	Type of rectifier circuit			
	Half-wave	Full-wave bridge	Centre-tap full-wave	Voltage doubler
(5) Conversion ratio $E_{dc}/E_{T(max)}$ using $\%$ R_S/R_L from (3) and the new value of $\omega R_L C$ from (4)	from Fig. 65 0·73	from Fig. 66 0·82	from Fig. 66 0·82	from Fig. 67 1·56
(6) $E_{T(max)} = \dfrac{E_{dc}}{\text{conversion ratio}}$	205	366	146·5	385
$E_{rms} = \dfrac{E_{T(max)}}{(2)^{\frac{1}{2}}}$ (V)	145	258	103·5	272
(7) Crest working voltage (V) that the rectifiers must with stand	205	366	293	770
(8) r.m.s. current per rectifier diode I_o	157	170	113	94
R_S/nR_L ($\%$)	6	3	3	4
I_{ms}/I_o	2·34	2·6	2·6	2·5
Average current (A) per rectifier diode I_o	1·5	1·0	1·0	1·0
therefore I_{ms} (A)	3·51	2·6	2·6	2·5
(9) Suitable rectifier diodes allowing for transient derating	BY × 38/600	BY × 38/1200	BY × 38/900	2 × BY × 38/1200
(10) Taking values of $n\omega R_L C$ and R_S/nR_L ($\%$) in (8) and using Fig. 70, I_{pk}/I_o	6·2	7·5	7·5	6·9
therefore I_{pk} (A)	9·3	7·5	7·5	6·9
(11) Initial switch-on current $I_{on} = E_{T(max)}/R_S$ (A) (check this with the surge current rating of the rectifier diodes selected)	34·2	40·7	41·2	32

TABLE 2 (*continued*)

Requirement	Type of rectifier circuit			
	Half-wave	Full-wave bridge	Centre-tap full-wave	Voltage doubler
(12) Transformer design for mains voltage of 230 V primary to secondary transformer ratio $N = 230/E_{T(rms)}$	1·585	0·892	2·22 (half secondary)	0·845
If primary winding resistance r_p (Ω) is	1·5	1·6	1·6	1·5
and secondary winding resistance r_s (Ω) is	2·0	2·0	1·2 (half secondary)	2·0
then transformer resistance referred to secondary is $r_s + (r_p/N^2)$ (Ω)	2·6	4·0	1·5	4·1
Voltage drop V_D (V) across rectifier at average current I_o	1·0	0·95	0·95	0·95
therefore rectifier resistance in circuit at average current is $r_r = V_D/I_o(\Omega)$	0·67	$2 \times 0·95$	0·95	$2 \times 0·95$
Total resistance in secondary circuit $r_s + (r_p/N^2) + r_r = r_{tot}$ (Ω)	2·27	5·9	2·45	6·0
External series resistance must be $R_S - r_{tot}$ (Ω)	2·73	2·1	1·15	6·0
Let $R_S - R_{tot}$ (Ω) be	2·0	2·0	1·0	6·0
Secondary r.m.s. current $I_{T(rms)}$ (A)	3·51	$(2)^{\frac{1}{2}} \times 2·6$ = 3·68	2·6 (half secondary)	$(2)^{\frac{1}{2}} \times 2·5$ = 2·55
Secondary r.m.s. voltage $E_{T(rms)}$ (V)	145	258	102·5	272
Secondary volt-ampere rating $E_{T(rms)} \times I_{T(rms)}$ (VA)	508	950	269 + 269	963
Power rating of series resistor (W)	27	40·6	12·6	75
(13) r.m.s. ripple current $I_{C(rms)}$ (A)	3·18 (Equation 23)	2·54 (Equation 24)	3·09 (Equation 24)	2·29 (Equation 23)

RC Damping Circuit

The *RC* damping circuit for the four examples in Table 2 can be designed by adopting the following procedure.   It may be connected to either the primary or the secondary of the transformer (see **BYX38** data).

The damping circuit components determined by using the expressions below are suitable for the suppression of transient voltages to less than $2V_{RW}$.

Consider the full-wave bridge rectifier circuit.   If the damping circuit is connected to the primary of the transformer, then

$$C_1 = 200\frac{I_{mag}}{V}\ \mu F \text{ and } R_1 = \frac{150}{C_1}\ \Omega,$$

where V is the transformer primary r.m.s. voltage and I_{mag} the magnetising primary r.m.s. current (A).

From Table 2 the primary r.m.s. current is $3 \cdot 68/0 \cdot 892$ A = $4 \cdot 13$ A. If $I_{mag} = 10\%$ of primary r.m.s. current, then

$$C_1 = 200\left(\frac{0 \cdot 413}{230}\right) = 0 \cdot 36\ \mu F.$$

Let $C_1 = 0 \cdot 5\ \mu F$; then

$$R_1 = \frac{150}{0 \cdot 5} = 300\ \Omega.$$

If the damping circuit is connected to the secondary,

$$C_2 = \frac{225(I_{mag}T^2)}{V}\ \mu F \text{ and } R_2 = \frac{200}{C_2}\ \Omega,$$

where

$$T = \frac{\text{transformer primary r.m.s. voltage}}{\text{transformer secondary r.m.s. voltage}}.$$

Therefore

$$C_2 = \frac{225 \times 0 \cdot 413}{230}\left(\frac{230}{258}\right)^2 = 0 \cdot 31\ \mu F.$$

Let $C_2 = 0 \cdot 5\ \mu F$; then $R_2 = 400\ \Omega$.

Heatsink Design

The heatsinks can be designed for the above four circuits from the information provided in the rectifier data.   The heatsink size for the bridge rectifier circuit is determined below to illustrate the procedure.

From the BYX38 data, for 1 A average current, a rectifier mounted on a heatsink with a thermal resistance of 14·4 degC/W $(R_{th(i)} + R_{th(h)}$ = 15 degC/W) can be operated up to an ambient temperature of 63 °C. If these fins are stacked to produce the bridge rectifier assembly, then the area of the heatsink should be approximately 30 cm^2 (one face).   A heatsink 6 cm × 6 cm should be satisfactory.

Performance

The voltage regulation curves for the four examples are shown in Figures 71 to 74.   From these curves it can be seen that the output voltage at the required current is within 2% of the stated value.

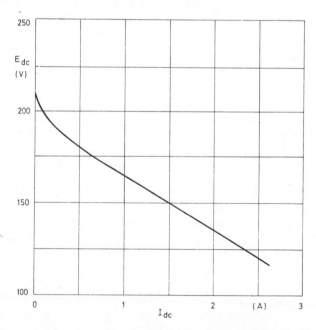

FIGURE 71.   Voltage regulation for single-phase half-wave circuit with capacitor input filter.

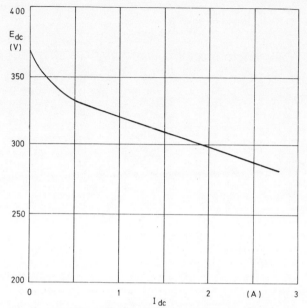

FIGURE 72.   Voltage regulation for single-phase full-
wave bridge circuit with capacitor input filter.

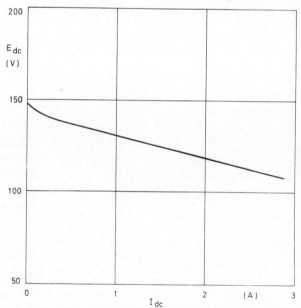

FIGURE 73.   Voltage regulation for single-phase centre-
tap circuit with capacitor input filter.

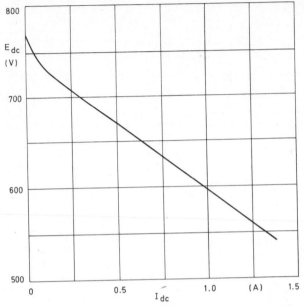

FIGURE   74.   Voltage   regulation   for   single - phase
voltage-doubler   circuit   with   capacitor   input   filter.

Design Procedure for Rectifier Circuits With Choke Input Filter

The analysis of the capacitor input filter rectifier circuits has shown
that, for any high-current conversion, the circuit requires a large value
of smoothing capacitor, which has to carry a large ripple current, and large
initial and repetitive peak currents flow through the rectifiers.   These
limitations may be overcome by the use of choke input filters.

The single-phase half-wave circuit (Figure 43) is not normally used with
a choke input filter, as it would require a high value of inductance to
cause current to flow throughout the cycle.

For the full-wave centre-tap circuit of Figure 47 and the full-wave bridge
circuit of Figure 49, R_s is replaced by a series choke L.

The full-wave bridge circuit with choke input filter is shown in Figure 75,
and the voltage and current waveforms in Figure 76.   The action of the
choke is to reduce both the peak and the r.m.s. value of current and to reduce
the ripple voltage.   The choke input filter circuit, however, requires a higher
applied voltage than the capacitor input filter circuit, to produce the same
output voltage.

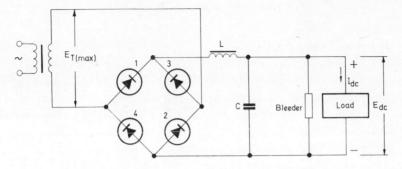

FIGURE 75. Single-phase full-wave bridge circuit with choke input filter.

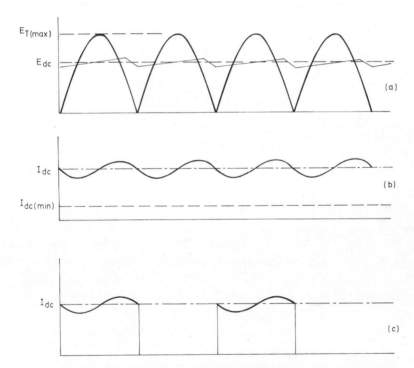

FIGURE 76. Waveforms for full-wave bridge circuit with choke input filter: (a) output voltage; (b) current through choke; (c) current through rectifiers (1 and 2) or (3 and 4).

Smoothing Circuit

The choke input filter should, ideally, pass only one frequency, which is zero, and attenuate all others. The filter should also allow direct current to flow to the load without much power loss, and at the same time present a high impedance to the fundamental and other ripple frequencies. To bypass the ripple current and its harmonics a capacitor shunting the load is used to complete the filter.

The attenuation factor K of the filter with series choke L and shunt capacitor C is defined as the ratio of the total input impedance of the filter to the impedance of the parallel combination of the shunt capacitor C and load R_L. For the choke input filter to function efficiently, the choke reactance at fundamental ripple frequency f_r should be much greater than its d.c. resistance, and the capacitor reactance much lower than the minimum load resistance.

If it is assumed that the inductance of the choke is L,

$$2\pi f_r L \gg \text{choke resistance } R_L$$

and

$$\frac{1}{2\pi f_r C} \ll R_{L(min)}$$

then

$$K = \frac{2\pi f_r L - 1/2\pi f_r C}{1/2\pi f_r C};$$

therefore

$$K = 4\pi^2 f_r^2 LC - 1. \tag{25}$$

The value of the inductance L used in the circuit must be such as to allow the rectifiers to conduct over one cycle of the fundamental ripple frequency. If the rectifier conducts for a period less than this, then the choke input filter will behave more like a capacitor input filter. This will give rise to a higher repetitive peak current through the rectifiers, and will also result in poor regulation.

The use of sufficient inductance allows the rectifier to conduct over the complete cycle, whereas the capacitor input filter allows the rectifier to conduct over only a fraction of a cycle. —It follows that, for a given load, the current will switch off just as the cycle is completed, for a certain value of inductance. This value is termed the critical inductance L_{crit}.

Output Voltage

Consider the single-phase full-wave bridge circuit shown in Figure 75 and the waveforms shown in Figure 76. The rectified voltage in Figure 76(a) is applied to the choke input filter. The crest value E_{max}, is equal to $E_{T(max)}$ in this circuit.

The rectified voltage can be approximated to a d.c. term plus a harmonic at the fundamental ripple frequency, if the amplitudes of the higher harmonics are negligible. Therefore,

$$e \approx \frac{2}{\pi} E_{max} - \frac{4}{3\pi} E_{max} \cos(2\omega t).$$

Critical Inductance

From Figure 76 it can be seen that, for the rectifier to conduct throughout the fundamental ripple cycle, the negative going peak ripple current delivered by the rectifier must not exceed the minimum d.c. current, which occurs with a load of $R_{L(max)}$. Thus

$$I_{dc(min)} = \frac{E_{dc}}{R_{L(max)}} = \frac{2E_{max}}{\pi} \frac{1}{R_{L(max)}} \qquad (26)$$

if $2\pi f_r L \gg R_L$, and $1/2\pi f_r C \ll R_{L(min)}$;

$$\text{peak ripple current} = \frac{4}{3\pi} E_{max} \frac{1}{2\pi f_r L}. \qquad (27)$$

The critical inductance is present when the peak a.c. equals the direct current. That is

$$\frac{4}{3\pi}E_{max}\frac{1}{2\pi f_r L_{crit}} = \frac{2E_{max}}{\pi}\frac{1}{R_{L(max)}};$$

therefore

$$L_{crit} = \frac{R_{L(max)}}{3\pi f_r}. \tag{28}$$

For 50 Hz supply frequency and full-wave rectification, $f_r = 100$ Hz, so that

$$L_{crit} = \frac{R_{L(max)}}{943}. \tag{29}$$

Because of the approximations made, it is necessary to use a somewhat higher value of inductance than L_{crit}. In practice, it is found that for reliable and satisfactory operation the optimum value of inductance which should be used is twice the value of L_{crit}.

It is obvious from the nature of the circuit that it is not possible to maintain the critical value of the inductance to low values of load current. This would require an infinite inductance at zero load current. Two methods are available to ensure that current flows throughout the cycle, and that good regulation is maintained over a wide range of load currents. These are the use of a bleeder resistance or a swinging choke.

Bleeder Resistance

A bleeder resistance of a suitable value is connected across the shunt capacitor to maintain the minimum current that will satisfy the critical inductance condition, even when no load is connected. The use of a bleeder will prevent the output voltage from rising to the peak applied voltage in the absence of the load.

Swinging Choke

The swinging choke method is based on the fact that the inductance of an iron-cored inductor partly depends on the amount of direct current flowing through it. The swinging choke is designed so that it has enough inductance at high currents, and this increases as the d.c. current is decreased. The use of such a choke is therefore very satisfactory for maintaining good regulation over a range of load current, and it is also more efficient than the bleeder resistance method.

The ripple voltage is no longer independent of the load current since the inductance is continually varying with the load. When using a swinging choke, it is necessary to ensure that the inductance does not fall too low at the maximum load current, as this will lead to high repetitive peak current and high ripple. In practice, the inductance at full load L_F should be such that

$$L_F = \frac{2R_{L(min)}}{943}.$$

Ripple Current and Voltage

If $2\pi f_r L \gg R_L$, $1/2\pi f_r C \ll R_{L(min)}$ and $2\pi f_r L \gg 1/2\pi f_r C$, then

$$\text{r.m.s. ripple current } I_{c(rms)} = \frac{4}{3} \frac{E_{max}}{\pi} \frac{1}{(2)^{\frac{1}{2}}} \frac{1}{2\pi f_r L}.$$

Since $E_{dc} = (2/\pi)E_{max}$,

$$I_{c(rms)} = \frac{(2)^{\frac{1}{2}}}{3} E_{dc} \frac{1}{2\pi f_r L}, \tag{30}$$

$$\% \text{ ripple} = \% \text{ ripple before filtering} \times \frac{1}{K}.$$

From Table 2, % ripple before filtering $= 47 \cdot 2\%$. From Equation (25), if $4\pi^2 f_r^2 LC \gg 1$, then $K \approx 4\pi^2 f_r^2 LC$ and

$$\% \text{ ripple} = \frac{47 \cdot 2}{4\pi^2 f_r^2 LC} = \frac{1 \cdot 193}{f_r^2 LC} \tag{31}$$

For 50 Hz supply frequency and full-wave rectification, $f_r = 100$ Hz; therefore

$$\% \text{ ripple} = \frac{119 \cdot 3}{LC},$$

where L is in henries and C is in microfarads.

Maximum Value of Shunt Capacitance

In evaluating the percentage ripple and the attenuation factor of the filter, it has been assumed that the reactance of the capacitor at the fundamental ripple frequency is very much lower than the minimum load resistance. In practice, it is found that satisfactory performance is obtained when the reactance of the capacitor is made less than one-fifth the minimum load resistance, that is

$$\frac{1}{2\pi f_r C} \leqslant \frac{R_{L(min)}}{5}. \tag{32}$$

Therefore

$$C \geqslant \frac{5 \times 10^6}{2\pi} \frac{1}{f_r R_{L(min)}} \mu F$$

$$\geqslant \frac{796\,000}{f_r R_{L(min)}}. \tag{33}$$

Because of the nature of the circuit, the capacitor will resonate with the inductor at a certain frequency. At this frequency the output impedance will be greater than the capacitor reactance. Therefore, when a variable loading is applied, precautions must be taken to ensure that the output impedance of the filter is small at the load current frequency.

Additional Filter Sections

When it is required to reduce the ripple voltage across the load to a very low value, a single-stage choke input filter may require large values of inductance and capacitance, which may lead to an uneconomic filter design. In this case, better results may be achieved by using a multi-stage filter with small value inductors and capacitors.   It can be shown that the optimum smoothing is achieved when all stages are identical   However a multistage filter attenuates low frequency transients less than a single stage filter.

Figure 77 shows the attenuation factor K plotted against $f_r^2 LC$ for one-stage, two-stage, and three-stage filters.   A suitable arrangement can therefore be selected by studying the filter characteristics.   For a K factor between 23 and 160, the two-stage filter is the most economic.   For K above 160 a three-stage filter is more suitable.

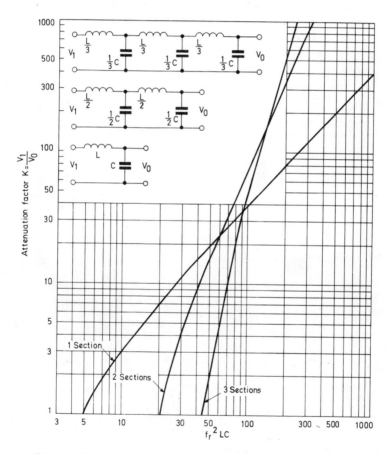

FIGURE 77.   Characteristics of choke input filters. L in hensies; C in microfarads.

Example on Full-wave Bridge Rectifier Circuit with Choke Input Filter

A full-wave bridge circuit to supply 0 to 4 A at 200 V is required.   The ripple is to be less than 0.5%, $f = 50$ Hz, choke resistance 7.5 Ω, and rectifier voltage drop approximately 1 V.

Let the bleeder current be 0.5 A; therefore the bleeder resistance R_b = $200/0.5$ = 400 Ω.   The external load at maximum current = $R_{L(min)}$ = $200/4$ = 50 Ω.   At zero load current, the total circuit resistance is approximately $400 + 7.5 = 407.5$ Ω.   From Equation (29) $L_{crit} = R_{L(max)}/943$ = $407.5/943$ = 0.432 H; therefore the optimum value is 0.864 H.

The relationships between a.c. and d.c. voltages and currents are given in Table 1 for circuits without filters.   The values given for inductive load circuits can also be used for choke input filter circuits.

In order to use the relationship of the idealised circuits (Table 1) E_{dc} must be increased above the output direct voltage required to allow for voltage drop across the choke and rectifiers.   That is

E_{dc} = direct output voltage required + volt drop across choke + volt drop across rectifiers.

Therefore

$$E_{dc} = 200 + 7.5(4 + 0.5) + (2 \times 1) \approx 236 \text{ V.}$$

From Table 1,
$$E_{rms} = 1.11E_{dc} = 262 \text{ V.}$$

From Equation (33)

$$C \geqslant \frac{796\,000}{f_r R_{L\,(min)}}.$$

For a bridge rectifier circuit, $f_r = 100$ Hz; therefore

$$C \geqslant \frac{796\,000}{100 \times 50} \geqslant 160 \text{ } \mu\text{F.}$$

From Equation (32), for ripple to be less than 0.5%

$$LC \geqslant \frac{119.3}{0.5}.$$

therefore

$$LC \geqslant 238\cdot6.$$

If $L = 1$ H, then $C \geqslant 238\cdot6\ \mu$F, so that a practical value of $C = 250\ \mu$F is suitable.

From Table 1, the crest working voltage that the rectifiers must withstand is $1\cdot57E_{dc} = 1\cdot57 \times 236 = 370$ V.

With allowance made for transients, BYX38 rectifier diodes should operate satisfactorily in this circuit.

From Table 1, $I_{pk} = I_{dc}$ for a pure inductive load. From Figure 76 it can be seen that I_{pk} is greater than I_{dc}, but not as large as the peak current as above with a resistive or capacitive load.

The maximum repetitive peak current per rectifier $I_{pk} < 1\cdot57I_{dc}$: therefore

$$I_{pk} < 1\cdot57 \times 4\cdot5 < 7\cdot6\ \text{A}.$$

The transformer rating can be determined by a similar procedure to that shown for the capacitor input filter rectifier circuits (page 96). The transformer resistance must be taken into account, and the transformer ratio determined accordingly to give 262 V r.m.s. on the secondary.

For a mains voltage of 230 V, primary winding resistance $r_p = 1\ \Omega$, and secondary winding resistance $r_s = 1\ \Omega$, the transformer ratio is

$$N = \frac{V_p}{V_s} = \frac{230}{262 + (r_s + r_p/N^2)I_{dc}} = 0\cdot843.$$

Secondary volt–ampere rating $= \dfrac{230}{0\cdot843} I_{dc} = 1230$ VA.

The RC damping circuit and heatsink must be designed according to the procedure given for the capacitor input filter rectifier circuits (page 99).

The voltage regulation curve for a circuit built with these components is shown in Figure 78, from which it can be seen that the output voltage at full load current is within 2% of the specified value. It can also be seen that the bleeder resistance is functioning correctly, since the current at which the voltage starts to rise rapidly is about one-half of the bleeder current.

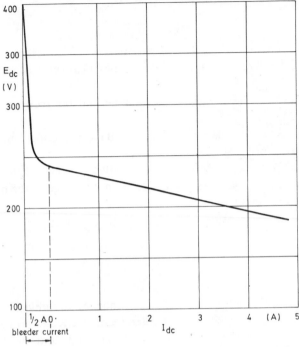

FIGURE 78. Voltage regulation for single-phase full-
wave bridge circuit with choke input filter.

Three- phase Rectifier Circuits

There are many advantages in using a polyphase rectifier system when
high-power conversion is required. The object is to superimpose more
voltages of the same peak value but in different time relation to each other.
An increase in the number of phases leads to. the following improvements:
(i) higher output voltage E_c for the same voltage input; (ii) higher fundamental ripple frequency and lower-amplitude ripple voltage; (iii) higher
overall efficiency.

For three-phase circuits, one winding of the transformer is generally
connected in delta to suppress harmonics (with the special exception of
the second double-star circuit on page 117). In the explanation of the
circuits in the next section, the secondary winding is always star-connected,
but delta connection could be used in the full-wave bridge circuit.

Half-wave, bridge, centre-tap, and double-star circuits will be discussed
in this section.

Three-phase Half-wave Circuit

The three-phase half-wave arrangement is the simplest three-phase rectifier circuit possible. It is shown in Figure 79. The secondary winding is star-connected, and the star point is used as a common load terminal. The relevant waveforms are shown in Figure 80.

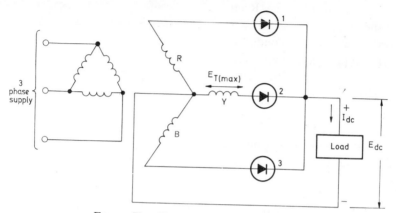

FIGURE 79.   Three-phase half-wave circuit.

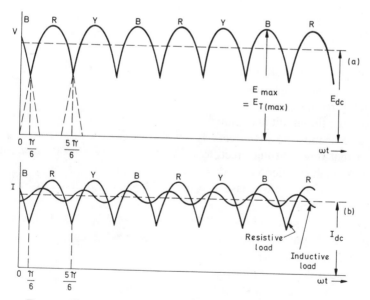

FIGURE 80.   Waveforms for three-phase half-wave circuit:
(a) voltage; (b) current.

The operation can best be understood by analysing the idealised wave-forms. Suppose that the voltage of phase R is most positive. Rectifier 1 will therefore conduct when $\omega t = \pi/6$, and the current will flow through the load and return to the transformer via the neutral point. Rectifier 1 will continue to conduct until the voltage of phase Y goes more positive than that of phase R at $\omega t = 5\pi/6$. The current will now be transferred from rectifier 1 to rectifier 2. Rectifier 2 will conduct for the next 120°, and then the current will be transferred to rectifier 3 for the next 120°. In this manner each rectifier conducts in turn for 120°.

The ripple frequency is three times the supply frequency, and the crest working voltage that the rectifiers must withstand is

$$2E_{T(max)}\cos(\pi/6) = (3)^{\frac{1}{2}}E_{T(max)}.$$

The conversion efficiency of this circuit is high in comparison with single-phase circuits, and the ripple voltage is reduced to little more than one-third of that obtained in the single-phase full-wave circuit. The transformer utility factor is, however, poor in comparison with the three-phase full-wave bridge rectifier, and the circuit is used only where low-voltage conversion is required.

Full-Wave Bridge Circuit

The three-phase full-wave bridge circuit is shown in Figure 81. It is one of the most widely-used circuits for high-power conversion with semi-conductor rectifiers.

Consider the circuit in conjunction with the waveform shown in Figure 82. If phase R is most positive, rectifier 1 will start conducting when $\omega t = \pi/6$. The current flows through rectifier 1 to the load and returns to the transformer through rectifier 5 or 6, depending on which phase, Y or B, is the more negative. At $\omega t = \pi/6$, phase Y is the most negative and therefore current will flow through rectifier 5. At $\omega t = \pi/2$ phase B goes more negative, and therefore current will now flow through rectifier 6 instead of rectifier 5. At $\omega t = 5\pi/6$, phase Y goes more positive and the current is therefore transferred from rectifier 1 to rectifier 2. Each rectifier conducts for 120° per cycle, and the current is commutated every 60°.

As in the single-phase bridge circuit (Figure 49), the crest working voltage, as given in Table 1, appears across two rectifiers. The ripple voltage is small, and the ripple frequency is six times the supply frequency.

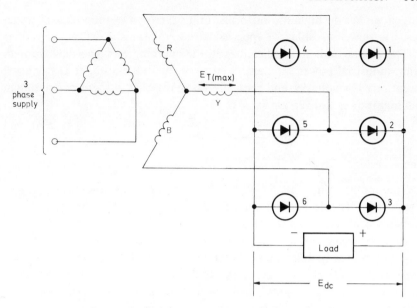

FIGURE 81.   Three-phase full-wave bridge circuit.

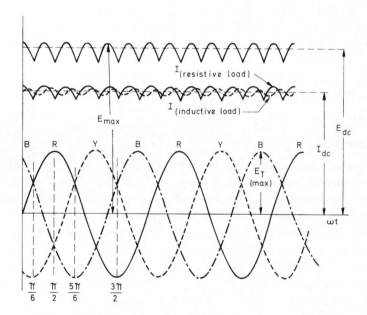

FIGURE 82.   Waveforms for three-phase full-wave bridge
circuit.

This circuit has the highest transformer utility factor, and it therefore requires least a.c. power to obtain a specified direct voltage and current.

The circuit finds its applications in the charging of higher-voltage batteries, industrial power supplies, electrolytic plant operating at any voltage except very low voltages, and generally where the most efficient and economical high-power conversion is required.

Double-bridge Circuit

The double-bridge circuit may be used where a very low ripple voltage is required.   The primary winding is either delta- or star-connected.   There are two sets of secondary windings.   One set is star-connected and the other is delta-connected.   Each set of windings is connected to a three-phase full-wave bridge rectifier assembly (Figure 81), and the output terminals of the two bridge circuits are connected in parallel.   If a threewire (centre earth) d.c. output is required, then the output terminals are connected in series.

The phase voltage of the secondary delta winding is $(3)^{\frac{1}{2}}$ times the phase voltage of the secondary star winding, so that the amplitudes of the output voltages from both the bridge rectifier circuits are the same.   However, the output voltage from the delta circuit is displaced in phase by $\pi/6$ relative to the output voltage from the star circuit.   The ripple frequency is therefore twelve times the supply frequency.   The percentage ripple is approximately 0.985%, and the output voltage $E_{dc} = 0.99E_{max}$ or $1.71E_{T(max)}$.

Centre-tap Circuit

The circuit for the three-phase centre-tap system, which is also known as a six-phase diametric circuit, is shown in Figure 83.   The centre tap on the transformer splits the three-phase supply to produce a six-phase supply.

The waveforms for this circuit are shown in Figure 84.

Each rectifier conducts for 60°, and the ripple frequency is six times the supply frequency.   This system has higher conversion efficiency than the half-wave three-phase circuit.   However, it has the lowest secondary utility factor of any three-phase circuit.   The conversion efficiency is high, and equal to that of the three-phase bridge.

The chief attraction of the circuit is that all rectifiers are connected to a common terminal, and therefore can be simply mounted on one heatsink.

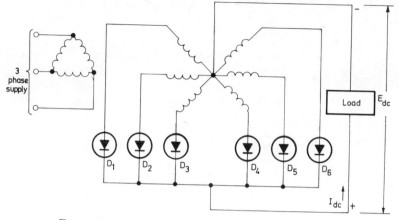

FIGURE 83. Three-phase full-wave centre-tap circuit.

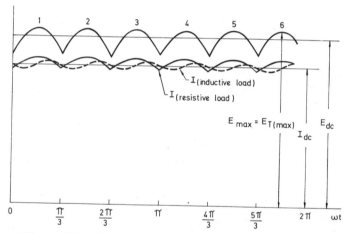

FIGURE 84. Waveforms for three-phase full-wave centre-tap circuit.

It is generally used for low-power conversion only, because of the poor secondary utility factor.

Double-star Circuit With Interphase Reactor

The double-star circuit with interphase reactor is shown in Figure 85. It has, in effect, two star-connected secondaries. The voltages of the two star connections are displaced by 180°. The neutral points of the two windings are connected together by a centre-tapped interphase reactor.

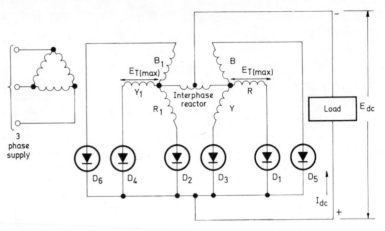

FIGURE 85.   Three-phase double-star circuit with interphase reactor.

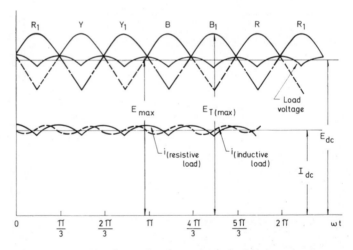

FIGURE 86.   Waveforms for three-phase double-star circuit
with interphase reactor.

At any instant, current is carried by two phases—one in each star—as shown in Figure 86.   The return current is divided between the two second-aries by the interphase reactor.   Thus the instantaneous output voltage is the average of the instantaneous voltages of the two secondaries that are conducting.   The variation in the d.c. current produces a third harmonic e.m.f. across each half of the interphase reactor, which adds to the e.m.f. of one anode and subtracts from that of the other, thus holding the two at a common voltage. At low d.c. currents, a transition point is reached when the current is too small to produce the third harmonic e.m.f. and the circuit reverts to the three-phase centre-tap system, giving a sudden rise in the output voltage.

The circuit has a six-phase ripple but a three-phase voltage ratio. Its use reduces the line current to approximately half that of the three-phase centre-tap circuit; therefore rectifiers with a smaller peak current rating can be used. However, the peak inverse voltage is somewhat greater.

This arrangement may be used where the cost of the interphase reactor is offset by the use of rectifiers with relatively low-current ratings. It is frequently used for low-voltage high-current electrolytic plant. The utility factors of both primary and secondary are high; the utility factor of the secondary is lower by a factor of $(2)^{\frac{1}{2}}$ than that of the three-phase bridge circuit.

Double-star Circuit Without Interphase Reactor

The purpose of the interphase reactor in the above circuit is to give a third harmonic e.m.f. which allows two rectifiers to conduct at the same time. A similar effect can be produced by using a transformer with a star-connected primary and a centre-tapped star secondary of the type shown in Figure 83. The two star points must not be connected.

With this arrangement, the transition from three-phase double-star operation to six-phase operation occurs at a higher current, unless special attention is given to the design of the transformer to give a high zero-phase sequence reactance. One method of obtaining this is to use a five-limb core with the windings on the three centre limbs.

The circuit is used for low-voltage electrolytic plant which is unlikely to be operated at currents less than 25% of full load current.

Smoothing of Three-phase Circuit Output

For the power visualised in three-phase silicon rectifier circuits, it would be prohibitive to use any filter circuit. A choke input filter may be a practical proposition where only a small current is required, but at high currents the size of the shunt capacitor required will be enormous, and it will have to carry a large ripple current.

For the sake of completeness, the value of critical inductance and various other relevant details have been given in Table 3. These values may be derived by a procedure similar to that described on page 105.

Table 3. Choke input filter performance

	Percentage ripple V_R (%)		Critical inductance L_{crit}(H)		R.M.S. Ripple current $I_{c(rms)}$(A)	
	General formula	50 Hz supply frequency	General formula	50 Hz supply frequency	General formula	50 Hz supply frequency
Single-phase full-wave	$\dfrac{1 \cdot 193}{f_r{}^2 LC}$	$\dfrac{119 \cdot 3}{LC}$	$\dfrac{R_{L(max)}}{3\pi f_r}$	$\dfrac{R_{L(max)}}{943}$	$\dfrac{E_{dc}}{13 \cdot 3 f_r L}$	$\dfrac{E_{dc}}{1330L}$
Three-phase half-wave	$\dfrac{0 \cdot 45}{f_r{}^2 LC}$	$\dfrac{20}{LC}$	$\dfrac{R_{L(max)}}{8\pi f_r}$	$\dfrac{R_{L(max)}}{3770}$	$\dfrac{E_{dc}}{35 \cdot 5 f_r L}$	$\dfrac{E_{dc}}{5310L}$
Three-phase full-wave bridge	$\dfrac{0 \cdot 102}{f_r{}^2 LC}$	$\dfrac{1 \cdot 133}{LC}$	$\dfrac{R_{L(max)}}{35\pi f_r}$	$\dfrac{R_{L(max)}}{33\,000}$	$\dfrac{E_{dc}}{155 f_r L}$	$\dfrac{E_{dc}}{46\,500L}$

$R_{L(max)}$ in ohms, C in microfarads, and L in henries.

Idealised Analysis of Polyphase Circuits

The voltage and current relationships, transformer rating, ripple, and interphase reactor rating for polyphase circuits are discussed in this section.

Voltage Relationships

The polyphase rectified output voltage from a sinusoidal supply can be represented by a series

$$e = E_{max}\frac{N}{\pi}\sin\left(\frac{\pi}{N}\right)\left\{1 + \frac{2\cos(N\theta)}{N^2 - 1} - \frac{2\cos(2N\theta)}{4N^2 - 1} + \frac{2\cos(3N\theta)}{9N^2 - 1} + \ldots\right\}, \qquad (34)$$

where E_{max} is the peak output voltage and N the number of output voltage pulses per cycle of supply voltage.

In the three-phase circuits discussed, N has the following values;

half-wave $N = 3$,
full-wave bridge $N = 6$,
centre-tap $N = 6$,
double-star $N = 6$.

The following analysis applies to all four of the three-phase circuits that have been discussed.   Modifications to the general analysis for particular circuit configurations are stated where necessary.

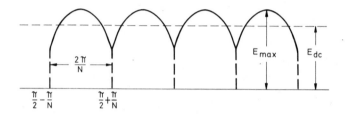

FIGURE 87.   Output voltage waveform fur three-phase
circuit.

Consider the output voltage waveform shown in Figure 87 where

$$E_{dc} = \frac{1}{2\pi/N} \int_{\pi/2 - \pi/N}^{\pi/2 + \pi/N} E_{max}\sin(\omega t) \, d(\omega t)$$

and therefore

$$E_{dc} = \frac{N}{\pi} E_{max}\sin\left(\frac{\pi}{N}\right)$$

The value of E_{dc} can also be directly obtained from Equation (34) where the first term represents E_{dc}.

The r.m.s. output voltage is given by

$$E_{rms} = \left\{ \frac{1}{2\pi/N} E_{max}^{2} \int_{\pi/2 - \pi/N}^{\pi/2 + \pi/N} \sin^2(\omega t) \, d(\omega t) \right\}^{\frac{1}{2}};$$

hence

$$E_{rms} = \frac{\pi}{N} \frac{E_{dc}}{\sin(\pi/N)} \left[\frac{N}{2\pi} \left\{ \frac{\pi}{N} + \frac{\sin(2\pi/N)}{2} \right\} \right]^{\frac{1}{2}}. \tag{35}$$

Current Relationships

For a resistive load,

$$I_{rms} \text{ per rectifier leg} = \frac{E_{rms}}{R} \frac{1}{(N)}.$$

Modifications are necessary for this equation to be valid for all four circuits. If I_{rms} is defined as

$$I_{rms} = \frac{E_{rms}}{R} \frac{1}{(N')},$$

then from Equation (35)

$$I_{rms} = \left(\frac{\pi}{N} \frac{I_{dc}}{\sin(\pi/N)} \left[\frac{N}{2\pi} \left\{ \frac{\pi}{N} + \frac{\sin(2\pi/N)}{2} \right\} \right]^{\frac{1}{2}} \right) \frac{1}{(N')^{\frac{1}{2}}}. \tag{37}$$

In this equation, the term $1/(N')^{\frac{1}{2}}$ has a denominator appropriate to the circuit configuration used. Thus, for three-phase half-wave $(N')^{\frac{1}{2}} = (3)^{\frac{1}{2}}$, for three-phase full-wave bridge $(N')^{\frac{1}{2}} = (3)$, for three-phase centre-tap $(N')^{\frac{1}{2}} = (6)^{\frac{1}{2}}$, and for three-phase double star $(N')^{\frac{1}{2}} = 2(3)^{\frac{1}{2}}$.

It should be noted that in the three-phase full-wave bridge circuit a pair of rectifiers conduct at any one instant; therefore $(N')^{\frac{1}{2}} = (3)^{\frac{1}{2}}$ and not $(6)^{\frac{1}{2}}$.

In the three-phase double-star circuit, the direct current is supplied by two separate star windings; therefore $(N')^{\frac{1}{2}} = 2(3)^{\frac{1}{2}}$.

For an inductive load,

$$I_{rms} \text{ per rectifier leg} = \frac{I_{dc}}{(N')^{\frac{1}{2}}}, \qquad (38)$$

where the values of $(N')^{\frac{1}{2}}$ given for Equation (37) also apply.

The average current per rectifier leg $I_o = I_{dc}/N$. This equation is valid for the three-phase rectifier circuits with the exception of the three-phase full-wave bridge. For this circuit, $N = 3$, because a pair of rectifiers conduct at any one instant.

Transformer Rating

The transformer secondary phase voltage $E_{T(rms)} = E_{max}(2)^{\frac{1}{2}}$. Modifications are necessary for this expression to be valid for all four circuits.

If $E_{T(rms)}$ is defined as

$$E_{T(rms)} = \frac{1}{K} \frac{E_{max}}{(2)^{\frac{1}{2}}},$$

then from Equation (34)

$$E_{T(rms)} = \frac{1}{K} \frac{\pi}{(2)^{\frac{1}{2}}N} \frac{1}{\sin(\pi/N)} E_{dc}, \qquad (39)$$

where for three-phase half-wave $K = 1$, for three-phase full-wave $K = (3)^{\frac{1}{2}}$, for three-phase centre-tap $K = 1$, and for three-phase double-star $K = (3)^{\frac{1}{2}}/2$.

For the three-phase full-wave bridge circuit, $K = (3)^{\frac{1}{2}}$ because the output voltage E_{max} is supplied by the three-phase line voltage.

For the three-phase double-star circuit, $K = (3)^{\frac{1}{2}}/2$ because

$$E_{max} = E_{T(max)}\cos(30^\circ) = \frac{(3)^{\frac{1}{2}}}{2} E_{T(max)}.$$

The transformer secondary r.m.s. current $I_{T(rms)} = I_{rms}$. This expression needs modification for the full-wave rectifier bridge circuit.

If $I_{T(rms)}$ is defined as

$$I_{T(rms)} = MI_{rms}, \qquad (40)$$

then for three-phase half-wave $M = 1$, for three-phase full-wave bridge $M = (2)^{\frac{1}{2}}$, for three-phase centre-tap $M = 1$, and for three-phase double-star $M = 1$.

For the three-phase bridge circuit, $M = (2)^{\frac{1}{2}}$ because each transformer winding supplies current to the circuit twice per cycle.

The secondary volt–ampere rating is

$$VA_s = n(E_{T(rms)}I_{T(rms)}), \tag{41}$$

where n is the number of secondary windings.

$$\text{Secondary utility factor} = \frac{E_{dc}I_{dc}}{VA_s} \tag{42}$$

Percentage Ripple

The percentage ripple is given by

$$\% \text{ ripple} = \frac{\text{fundamental r.m.s. ripple voltage}}{E_{dc}} \times 100.$$

From Equation (34), if ripple frequencies other than the fundamental are ignored,

$$\% \text{ ripple} = \frac{2}{N^2 - 1} \frac{1}{(2)^{\frac{1}{2}}} \times 100 = \frac{141}{N^2 - 1}. \tag{43}$$

The fundamental ripple frequency is

$$f_r = Nf, \tag{44}$$

where f is the supply frequency.

Rating of Interphase Reactor

Rectification with a double-star circuit requires an interphase reactor. The rating of this reactor can be calculated as follows.

If it is assumed that a triangular waveform appears across the reactor in the process of holding the phase voltages of the two star circuits at a common value, the crest value of this waveform will be V_{max}.   The frequency is three times the supply frequency.   The voltage across the reactor is at its maximum when the phase voltage of one star connection is displaced by $\pi/3$, and therefore is at half-maximum when the maximum voltage appears across the reactor.   Thus

$$V_{max} = E_{T(max)} - \frac{E_{T(max)}}{2}$$

$$= \frac{E_{T(max)}}{2} = \frac{(2)^{\frac{1}{2}}}{2} E_{T(rms)}$$

The triangular waveform can also be represented by a sine series

$$v = V_{max} \frac{8}{\pi^2} \left(\sin(\theta) - \frac{1}{9} \sin(3\theta) + \frac{1}{25} \sin(5\theta) - \cdots \right). \qquad (45)$$

If the third and higher harmonics are ignored, the peak value of an equivalent sine wave is $E_{eq(max)}$.   Thus

$$E_{eq(max)} = \frac{8}{\pi^2} V_{max} = \frac{8}{\pi^2} \frac{(2)^{\frac{1}{2}}}{2} E_{T(rms)};$$

therefore

$$E_{eq(rms)} = \frac{1}{(2)^{\frac{1}{2}}} \left\{ \frac{8}{\pi^2} \frac{(2)^{\frac{1}{2}}}{2} E_{T(rms)} \right\} = \frac{4}{\pi^2} E_{T(rms)}.$$

Now the form factor for a triangular waveform is r.m.s./mean $= 1 \cdot 16$; therefore the r.m.s. voltage rating of the reactor is

$$\frac{4}{\pi^2} \frac{E_{dc}}{1 \cdot 16}.$$

The current through the reactor is $I_{dc}/2$; therefore the power rating of the reactor is

$$\frac{4}{\pi^2 \times 1\cdot16 \times 2} E_{dc}I_{dc} = 0\cdot174E_{dc}I_{dc}. \tag{46}$$

Comparison of Three-phase Circuit Performances

Table 1 includes the performance of the commonly used three-phase rectifier circuits.   In evaluating the results in this table, it has been assumed that the transformer and rectifiers are ideal.   The table, however, gives a good indication of the relative merits of the circuits, and may be used to select the best circuit for any particular application.   It may also be used for comparing the kilowatts per rectifier available from various circuits. This is best illustrated by an example.

Consider the single-phase and three-phase full-wave bridge circuits, with rectifiers rated at a crest working voltage of 400 V and with a current rating of 20 A.   The attainable performances are compared in Table 4.

From the above calculation, it follows that a better use of rectifiers is made in the three-phase bridge circuit.

Table 4.   Comparison of three-phase circuits

	Single-phase bridge	Three-phase bridge
Number of rectifiers in circuit from Table 1	4	6
Output voltage E_{dc} (V)	$\frac{400}{1\cdot57} = 255$	$\frac{400}{1\cdot05} = 380$
Output current I_{dc} (A)	$2 \times 20 = 40$	$3 \times 20 = 60$
Power available $E_{dc}I_{dc}$	$0\cdot255 \times 40 = 10\cdot2$	$0\cdot380 \times 60 = 22\cdot8$
Kilowatts per rectifier	$\frac{10\cdot2}{4} = 2\cdot55$	$\frac{22\cdot8}{6} = 3\cdot8$

Idealised Analysis of Three-phase Bridge Circuit

The three-phase bridge circuit will be analysed as an example of how various values tabulated in Table 1, are determined.

The output voltage waveform is given by Equation (34). The number N of voltage pulses per cycle of mains voltage is 6. Therefore

$$e = \frac{3}{\pi} E_{max} \left(1 + \frac{2}{35} \cos(6\theta) - \frac{2}{143} \cos(12\theta) + \ldots \right)$$

and

$$E_{dc} = \frac{3}{\pi} E_{max} = 0.955 E_{max}$$

or

$$E_{max} = \frac{E_{dc}}{0.955} = 1.05 E_{dc}.$$

From Equation (35)

$$E_{rms} = \frac{\pi}{3} E_{dc} \left[\frac{6}{2\pi} \left\{ \frac{\pi}{6} + \frac{(3)^{\frac{1}{2}}}{4} \right\} \right]^{\frac{1}{2}} :$$

therefore

$$E_{rms} = 1.0 E_{dc} .$$

The average output current per rectifier leg $= I_{dc}/3 = 0.33 I_{dc}$. The r.m.s. current per rectifier leg for resistive load is, from Equation (36),

$$I_{rms} = \frac{E_{rms}}{R} \frac{1}{(N')^{\frac{1}{2}}} = \frac{E_{dc}}{R} \frac{1}{(N')^{\frac{1}{2}}} = \frac{1}{(3)^{\frac{1}{2}}};$$

therefore

$$I_{rms} = 0.577 I_{dc} .$$

With an inductive load

$$I_{rms} = \frac{I_{dc}}{(N')^{\frac{1}{2}}} = 0.577 I_{dc}.$$

The peak current per rectifier leg for a resistive load is

$$I_{pk} = \frac{E_{max}}{R} = \frac{1 \cdot 05 E_{dc}}{R} = 1 \cdot 05 I_{dc},$$

and for an inductive load

$$I_{pk} = I_{dc}.$$

The transformer secondary r.m.s. current is, from Equation (40),

$$I_{T(rms)} = (2)^{\frac{1}{2}} I_{rms} = 0 \cdot 816 I_{dc}.$$

The transformer secondary r.m.s. voltage is, from Equation (39),

$$E_{T(rms)} = \frac{1}{(3)^{\frac{1}{2}}} \frac{\pi}{6(2)^{\frac{1}{2}}} 2E_{dc} = 0 \cdot 428 E_{dc} \,;$$

therefore the total transformer secondary volt-ampere rating is, from Equation (41),

$$\begin{aligned} VA_s &= (E_{T(rms)} \times I_{T(rms)}) \times 3 \\ &= (0 \cdot 428 \times 0 \cdot 816) \times 3 E_{dc} I_{dc}. \\ &= 1 \cdot 05 E_{dc} I_{dc}. \end{aligned}$$

The secondary utility factor, from Equation (42), is

$$U_s = \frac{E_{dc} I_{dc}}{1 \cdot 05 E_{dc} I_{dc}} = 0 \cdot 95.$$

Since the line current is symmetrical in this circuit,

primary r.m.s. current.=. secondary r.m.s. phase current times turns ratio

$$= 0 \cdot 816 I_{dc} \frac{N_s}{N_p}.$$

Primary r.m.s. phase voltage $= E_{T(rms)} \dfrac{N_p}{N_s}$

$$= 0 \cdot 428 E_{dc} \frac{N_p}{N_s};$$

therefore

primary volt-ampere rating $VA_p = 0 \cdot 816 I_{dc} \dfrac{N_s}{N_p} \times 0 \cdot 428 E_{dc} \dfrac{N_p}{N_s}$

$$= 1 \cdot 05 E_{dc} I_{dc}.$$

$$\text{Primary utility factor} = \frac{E_{dc} I_{dc}}{1 \cdot 05 E_{dc} I_{dc}}.$$

Fundamental ripple frequency $= 6f$ (from Equation (44)).

Percentage ripple $V_R(^\circ/_\circ) = \dfrac{141}{6^2 - 1} = 4 \cdot 03^\circ/_\circ$ from Equation (43)).

Crest working voltage $= 2 E_{T(max)} \cos \dfrac{5\pi}{6}$

$$= 2(2)^{\frac{1}{2}} \frac{(3)^{\frac{1}{2}}}{2} E_{T(rms)} = 2 \cdot 45 E_{T(rms)}$$

and

crest working voltage $= 2 \cdot 45 \times 0 \cdot 428 E_{dc} = 1 \cdot 05 E_{dc}.$

Losses in Three-phase Circuits

The output voltage in a practical rectifier circuit is lower than the ideal value because of regulation. The voltage regulation of the three-phase system depends on three factors; the copper loss of the transformer, the rectifier voltage drop, and the commutation voltage drop.

Copper Loss

The reduction in output voltage due to transformer copper loss can be calculated as follows:

voltage drop due to copper loss

$$E_K = \frac{\text{transformer copper loss in watts } P_K}{I_{dc}} \qquad (47)$$

The value of P_K may be obtained from the short-circuit test on the transformer (Ref. 107).

Rectifier Forward Voltage Drop

The loss due to the forward voltage drop of the rectifier is generally small, especially with silicon rectifiers, which have a voltage drop of only one or two volts.   An accurate value for any particular type can be obtained from the forward voltage–forward current characteristic of the rectifier.

The effect of this loss will depend on how many rectifiers are used in series.   In particular, it should be noted that, in any bridge circuit with one rectifier per leg, the forward voltage drop is that due to two rectifiers in series.

Commutation Loss

The inductance of the transformer winding prevents the current from transferring instantaneously from one phase to the next.   Thus for a period the two rectifiers conduct simultaneously.   During the commutation period the rectifier output voltage is the average of the instantaneous voltages of the two phases; therefore the output voltage is reduced by the shaded area shown in Figure 88.

During the commutation period the d.c. current is the sum of the increasing current of the oncoming rectifier diode and the decreasing current of the previously conducting rectifier diode.   The commutation period ends when the current flowing through the conducting rectifier diode falls to zero, since it cannot pass any current in the reverse direction.

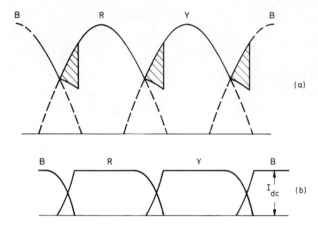

FIGURE 88.　Commutation loss due to transformer
reactance: (a) voltage; (b) current.

The voltage drop due to commutation. E_{com}, increases with an increase
in the number of phases and an increase in the load current. In order to
keep the commutation loss to a minimum, the commutating reactance must
be limited to a small value, bearing in mind that, under short-circuit conditions, the short-circuit current which is limited by the commutating
reactance must not exceed the surge current rating of the rectifier diode.
The commutation loss is given by

$$E_{com} = \frac{nX_L I_{dc}}{2\pi},\qquad(48)$$

where n is the number of phases, X_L the transformer reactance per phase,
and I_{dc} the d.c. load current.

With all these losses taken into account, the direct voltage appearing
across the load is $E_{dc(actual)}$;

$$E_{dc(actual)} = E_{dc(idealised)} - \frac{nX_L I_{dc}}{2\pi} - \frac{P_k}{I_{dc}} - V_D X,\qquad(49)$$

where V_D is the forward voltage drop per rectifier diode and X is the number
of rectifiers in series.

Examples of Three-phase Rectifier Circuits

Consider a three-phase rectifier circuit to supply 90 A. Suppose the available rectifier diodes have a maximum crest working voltage rating of 400 V. The transformer has a percentage reactance of 5% and a copper loss of 900 W.

The design of suitable bridge and centre-tap circuits is summarised in Table 5.

Table 5. Design of three-phase circuits

	Three-phase bridge	Three-phase centre-tap
From Table 1		
Maximum a.c. r.m.s. voltage that may be applied (V)	$\dfrac{400}{2 \cdot 45} = 163$	$\dfrac{400}{2 \cdot 83} = 141 \cdot 4$
Idealised output direct voltage (V)	$\dfrac{163}{0 \cdot 428} = 380$	$\dfrac{141 \cdot 4}{0 \cdot 74} = 191 \cdot 2$
Average current per rectifier leg (A)	$\dfrac{90}{3} = 30$	$\dfrac{90}{6} = 15$
Suitable rectifier diode	BYY15	BYX13-800
Total diode voltage drop at average current (V)	$0 \cdot 98 \times 2 = 1 \cdot 96$	$0 \cdot 96$
Voltage drop due to commutation overlap (Equation (48)) (V)	$\dfrac{3 \times 0 \cdot 05 \times 163}{2\pi} = 3 \cdot 9$	$\dfrac{6 \times 0 \cdot 05 \times 141 \cdot 4}{2\pi} = 6 \cdot 75$
Voltage drop due to copper loss (V)	$\dfrac{900}{90} = 10$	$\dfrac{900}{90} = 10$
Approximate voltage available at output terminals (V)	$380 - 1 \cdot 96 - 3 \cdot 9 - 10 = 364$	$191 \cdot 2 - 0 \cdot 96 - 6 \cdot 75 - 10 = 173$
Output power $E_{dc} I_{dc}$ (kW)	$0 \cdot 364 \times 90 = 32 \cdot 8$	$0 \cdot 173 \times 90 = 15 \cdot 6$

Transformer Reactance and Circuit Efficiency

The regulation of the three-phase rectifier circuits depends mainly on the transformer performance. In order to estimate the transformer performance, it is necessary to carry out an open-circuit and short-circuit test on the transformer. The purpose of each is briefly outlined below; but, for details of these tests, appropriate literature should be consulted (Ref. 107).

Open-circuit Test

With either the primary or the secondary open-circuited, the current and power at normal voltage and frequency are measured. The current $I_{o/c}$ is the sum of the magnetising current and core loss components. The power indicated, $W_{o/c}$, represents the core loss and copper loss. The latter is small and therefore may be neglected, since $I_{o/c}$ is small compared with the full load current.

Short-circuit Test

In the short-circuit test, either primary or secondary is short-circuited, and the voltage is gradually increased to circulate the rated current through the winding. The short-circuit voltage $V_{s/c}$ necessary to circulate the full load current is measured. The power reading $W_{s/c}$ in this test represents the copper loss I^2R and a small core loss that may be ignored.

Calculation

From the above two tests, the performance of the transformer may be calculated as follows.

> Transformer rating $= M$ volt ampere.
> Transformer connection $=$ delta-star.
> Normal primary voltage $= E_p$ volts
> Normal secondary voltage $= E_s$ volts

Open circuit test on star side at normal voltage E_s:
> Core loss $= W_{o/c}$,
> No-load current $= I_{o/c}$.

Short-circuit test on delta side with secondary short-circuited,

Short-circuit voltage $= V_{s/c}$.

Copper loss $P_K = W_{s/c}$ watts at rated current.

Primary line current $= I_p = \dfrac{M}{(3)^{\frac{1}{2}} E'_p}$

Primary phase current $= I_p/(3)^{\frac{1}{2}}$.

Copper loss per phase $= \dfrac{W_{s/c}}{3}$ watts.

Reactance e.m.f. per phase at current $I_p/(3)^{\frac{1}{2}} = E_x$

$$E_x = \left[V_{s.c}^{\;2} = \left\{ \frac{W_{s/c}}{3} \frac{(3)^{\frac{1}{2}}}{I_p} \right\}^2 \right]^{\frac{1}{2}} \tag{50}$$

Therefore % reactance $= \dfrac{E_x}{E_p} 100 = X\%$.

Circuit Efficiency

$$\% \text{ efficiency} = \frac{\text{output}}{\text{output} + \text{losses}} 100$$

$$= \left(1 - \frac{\text{losses}}{\text{output} + \text{losses}} \right) \times 100, \tag{51}$$

where total losses $= W_{o/c} = W_{s/c} + I_{dc} V_D \times$ number of rectifiers in series.

Rectifier Diodes at High Frequencies (kHz)

Published data sheets on rectifier diodes usually give limiting values that apply for sinusoidal operation at frequencies from 50 Hz to 400 Hz. The performance of rectifier circuits at these frequencies has already been described in previous sections. At higher frequencies, reverse recovery transients contribute considerably to the heating of the diode junction. It is therefore necessary to take into account the dissipation caused by these transients so that the operating temperature is kept within safe limits.

In this section the reverse recovery phenomenon and its influence on the high-frequency performance of rectifier diodes is investigated.   The minority carrier storage curves for sinusoidal and square-wave operation are obtained; and it is shown that the minority carrier storage charge for sinusoidal operation can be estimated, if the transient behaviour of the rectifier during square-wave operation is known.   From the experimental results the dissipation due to reverse recovery effects is estimated.   It is shown how current derating curves can be derived for square-wave, pulse, and sinusoidal operation of diodes at high frequencies.   The beneficial effect of improved cooling is demonstrated   The rectification efficiencies are also calculated.

The work described in this section was performed on devices similar to BYX38 diodes (Ref. 12).   This type was used for experimental convenience.   The results should therefore be regarded as a statement of a general problem rather than as an examination of BYX38 performance at frequencies above the published maximum.

The devices used here have been uprated since the completion of the work described.   Designers should therefore base their calculations on the current data sheet for a device under consideration, and should regard the curves and calculations in this section as illustrative examples only.

Square-wave Operation at High Frequencies

When the rectifier diode which is conducting has reached a steady state, there is always a minority carrier charge present corresponding to the current flowing through the diode.   The minority carriers are continuously recombining and new carriers taking their place.   The average time of recombination depends on the parameters of the device.

The charge extracted under specified switching conditions will be independent of frequency, provided that the pulse duration is sufficient to allow the diode to attain its steady state.   The dissipation caused by the reverse transient is directly proportional to the repetition rate of the transient, that is to the operating frequency.

Reverse Switching Transient Measurement

The reverse recovery transients for various currents can be measured for the condition where the reverse current is equal to the forward current. These values can be used in order to establish a standard for measurements.

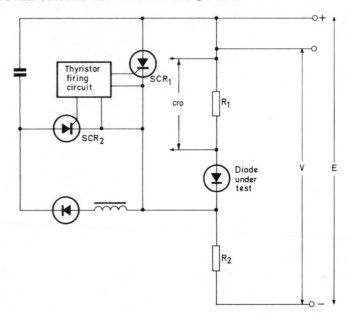

Figure 89.   Measurement of minority carrier charge under
pulse conditions.

The circuit for measuring the extracted charge is shown in Figure 89. The thyristor circuit is a conventional cathode pulse turn-off circuit (Ref. 33). It is used here to apply a switching voltage to reverse-bias the diode under test.

The rise time of SCR_1, which is switched to apply the reverse voltage, must be fast, and the gate pulse applied must be such as to allow SCR_1 to switch on at this rate.   Connecting leads should be as short as possible to minimise any inductance in the circuit.

The forward current I_{pk} through the diode under test is set by the applied voltage V and the non inductive resistors R_1 and R_2.   Thus

$$I_{pk} = \frac{V - V_F}{R_1 + R_2}.$$ (52)

When SCR_1 is turned on, the applied voltage E should be such that, with the diode under test short-circuited, the reverse current through resistor

R_1 is equal to the forward current I_{pk}. Thus

$$I_R = I_{pk} = \frac{E - E_D - V}{R_1},$$

where E_D is the forward voltage drop across SCR_1; therefore

$$E = V + E_D + I_{pk}R_1. \tag{53}$$

Typical current waveforms observed across R_1 are shown in Figure 90, and the shaded areas represent the extracted charge. At low currents the reverse transient waveform is of the shape shown in Figure 90(a). It resembles the idealised waveform (Figure 6). At high currents the rectifier starts to recover before the full reverse current can pass through it, and the shape of the transient is as shown in Figure 90(b).

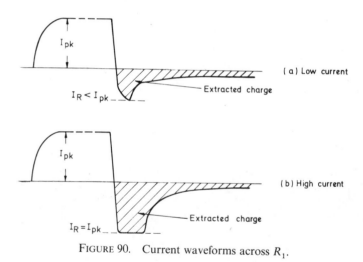

FIGURE 90.   Current waveforms across R_1.

The rise time of SCR_1 was such as to force the current through the rectifier at the rate of 3 A/μs. Figure 91 shows the extracted charge against forward current curves for three BYX38 rectifier diodes. These curves represent the typical, maximum, and minimum extracted charge measured

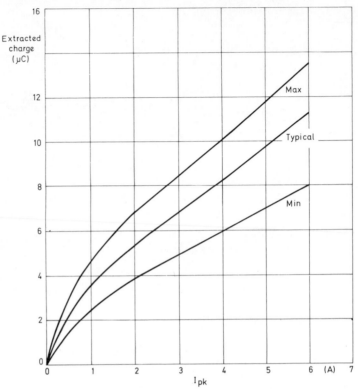

FIGURE 91.   Extracted charge as a function of peak current for three BYX38 with minimum, typical, and maximum extracted charge.

on some samples available at the time of this investigation.   All further considerations in this section are based on the sample from this batch with the maximum extracted charge, since this is the worst condition for high-frequency operation.

For BYX38 rectifier diodes, the extracted charge is mainly the charge due to minority carriers, since the rectifier capacitance is small.   The capacitance of a BYX38 rectifier diode at 30 MHz was found to be about 6 pF at 400 V.

Effect of Delay on Extracted Charge

When a rectifier diode is switched from a steady forward current I_{pk} to zero current with no reverse bias, the minority carrier charge within the device,

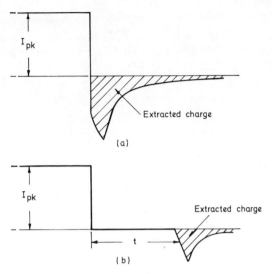

FIGURE 92.   Influence of delay t on extracted
charge.

corresponding to this peak current, reduces itself by the process of recombination.   If, therefore, the reverse bias is applied after a delay t, as shown
in Figure 92(b), then the extracted charge will be less than if the reverse
bias were applied with no delay as shown in Figure 92(a).   The difference
between the two extracted charges represents the charge that has recombined
during the delay t.

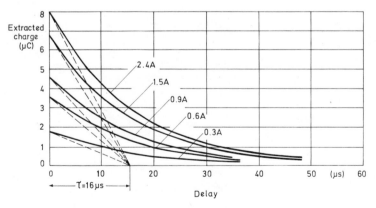

FIGURE 93.   Extracted charge as a function of delay, for various
peak currents.

Figure 93 shows the effect of delay on the extracted charge for various forward currents, as measured on a BYX38. The decay is exponential, and the time constant of the decay ($\tau = 16\mu s$) for all five curves is approximately constant.

An estimate of the initial charge can now be made   If the current through the rectifier diode reverses at a rate such that it passes through zero current at time t_1 (which is short compared with τ), then during this period the charge is given by the following equation:

$$Q = Q_i \exp\left(-\frac{t}{\tau}\right), \qquad (54)$$

where $t \leqslant t_1$, Q_i is the initial charge in the device corresponding to I_{pk}, and Q the charge remaining in the device after time t.

As soon as reverse current flows, the process of reduction of charge becomes complex. Part of the charge is being extracted during the period t_{rr}, and at the same time part of the charge is recombining. Figure 94 shows the estimated initial charge Q_i, and the extracted and recombined charges. The estimated initial charge is less than the actual initial charge by the amount that has recombined during the period t_{rr}. Figure 95 shows the extracted charge and the estimated initial charge Q_i against forward current I_{pk} plotted according to Equation (54), for the 'maximum' curve of Figure 91 and $\tau = 16 \ \mu s$.

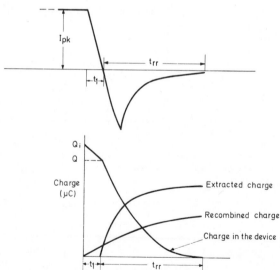

FIGURE 94.   Actual charge in rectifier diode.

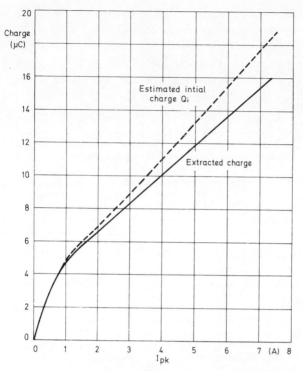

FIGURE 95. Extracted charge and estimated initial
charge as functions of peak current.

Dissipation due to Reverse Switching Transients

The determination of the exact amount of dissipation caused by reverse recovery transients is difficult for a number of reasons. Several parameters of the rectifier diode are functions of the temperature and geometry of the device. In addition, the amount of charge extracted will depend on the individual rectifier diode, the circuit impedance, and the amplitude and rate of the switching voltage. The shape of the switching transient is therefore likely to vary from one rectifier diode to another, and also from one type of rectifier diode to another.

For the samples of BYX38 rectifier diode investigated, it was observed that the recovery time and the extracted charge under specified conditions $(I_R = I_{pk})$ increased with increasing temperature. The extracted charge reached a nearly constant value at a case temperature above 60 °C.

The reverse dissipation loss due to the reverse switching transient was estimated for the condition where the reverse current is restricted to the same value as the forward current by circuit conditions. This part of the investigation was carried out on a BYX38 sample that exhibited the worst high-frequency characteristic but was still within the ratings specified in the published data. Thus a rectifier diode of the same type but with a better high-frequency characteristic will have less dissipation loss than that estimated for the worst device.

Estimation of Dissipation

Several assumptions were made in the determination of the dissipation loss due to reverse switching transients. Is was assumed that during the period t_2 (Figure 96) the diode junction is in a short-circuit condition, and

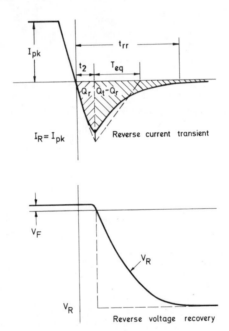

FIGURE 96. Approximation for estimating reverse transient dissipation, for square-wave operation.

therefore the peak reverse current flowing through the junction at time t_2 is $I_R = I_{pk}$. The dissipation loss during this period is small (since the junction is still forward-biased) compared with the dissipation loss during the remainder of the recovery period.

It was observed from measurements on several samples of BYX38 that the charge Q_r (Figure 96) was extracted before the diode resistance began to build up rapidly to its high value. The ratio of Q_r to the total charge extracted Q_t for any one rectifier diode remained approximately constant for all values of forward current. The value of this ratio, however, varies from one rectifier diode to another. For the samples investigated, the spread was from 0·3 to 0·45. The lower value corresponds to the device with the worst high-frequency characteristic and therefore the maximum dissipation.

For the purpose of estimation of the loss, several approximations to the transient waveforms were tried. The following method proved to be the most satisfactory. The approximation for this method is shown in Figure 96 by dashed lines.

Instead of using the experimental peak reverse current, it was assumed that it reached $I_R = I_{pk}$. The reverse current was assumed to decay linearly from $I_R = I_{pk}$ to zero in time T_{eq}, that is the transient was assumed to be triangular and to enclose the same area as the reverse switching transient during the period $t_{rr} - t_2$. The current waveform can then be defined by the following two equations:

$$\frac{I_R T_{ep}}{2} = Q_t - Q_r. \tag{55}$$

During the period T_{eq}, the instantaneous current i_R at time t (where $T_{eq} + t_2 \geqslant t \geqslant t_2$) is given by

$$i_R = I_R \left(1 - \frac{t - t_2}{T_{eq}} \right). \tag{56}$$

The shape of the actual recovery voltage waveform is shown by the solid line in the voltage waveform of Figure 96. As a simplification, the voltage was assumed to rise instantaneously at time t_2 to the peak applied reverse voltage V_R, so that

$$v_R = V_R. \tag{57}$$

The dissipation loss W_R at frequency f is given by the following expressions:

$$W_R = f \int_{t_2}^{T_{eq} = t_2} v_R i_R \, dt, \qquad (58)$$

from which

$$W_R = \frac{I_R V_R T_{eq} f}{2}. \qquad (59)$$

The dissipation loss estimated by means of the above equation gave results that were 20 to 30% higher than the experimental values. . However, the error is in the right direction, and therefore gives an added margin of safety.

To generalise the procedure, let

$$Q_r = k Q_t. \qquad (60)$$

From Equation (55) and (60)

$$W_R = (1 - k) Q_t V_R f. \qquad (61)$$

For the devices investigated, the lowest value of k determined was $0 \cdot 3$. Making allowances for temperature variation and so on, let $k = 0 \cdot 2$. Then

$$W_R = 0 \cdot 8 Q_t V_R f. \qquad (62)$$

Curves for W_R against frequency, for a fixed voltage of $V_R = 100$ V and for various values of forward current, are shown in Figure 97. The value of $Q_t = Q_i$ obtained from Figure 95 was used.

This graph has been drawn for $I_R = I_{pk}$, but it is also applicable for any value of I_R, because Q_t remains unaltered.

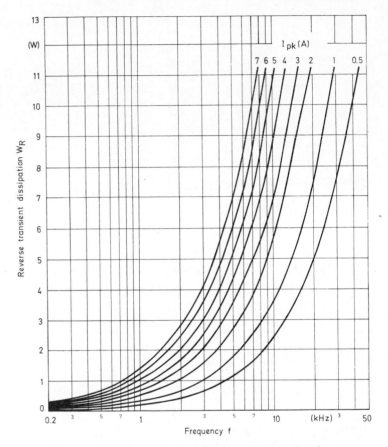

FIGURE 97.   Dissipation as a function of operating frequency, for reverse voltage V_R of 100 V, at various peak currents.   For dissipation at any other reverse voltage V_{R1}, multiply the value read off by $V_{R1}/100$.

Derating of Rectifier Diodes

It is seen from Equation (62) that the dissipation loss of the rectifier diode due to reverse switching transients increases with frequency.   If the frequency is high enough, these losses are substantial and must be taken into account. Either the cooling facilities must be improved, or the rectifier current must be derated to ensure that the additional dissipation does not damage the rectifier diode.

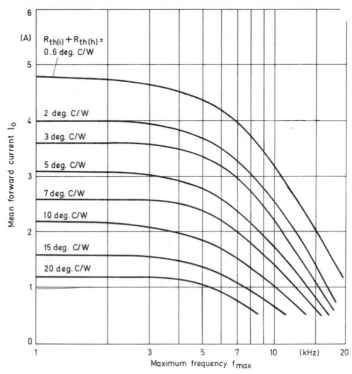

FIGURE 98. Current derating for square-wave operation, with reverse voltage V_R of 100 V, maximum ambient temperature of 55 °C, and various values of $R_{th(i)} + R_{th(h)}$. For f_{max} at any other reverse voltage V_{R1}, multiply the value read off by $100/V_{R1}$.

Figure 98 shows the derating curves for square-wave operation up to an ambient temperature of 55 °C.

The derating curves of Figure 98 were derived from the design curves given in the BYX38 published data extant at the time of the experimental work, and the dissipation loss calculated from Equation (62) or obtained from Figure 97. The procedure can best be illustrated by an example.

Suppose that a BYX38 rectifier diode is mounted on a heatsink with a thermal resistance $R_{th(h)}$ of 4·4 degC/W. The supply voltage is a 200 V (peak to peak) square wave, and the maximum ambient temperature is 55 °C. What is the maximum frequency at which a mean current of 1 A can be passed through the rectifier?

From the published data, for an ambient temperature of 55 °C, and low-frequency operation $R_{th(h)} + R_{th(i)} = 4\cdot4 + 0\cdot6 = 5$ degC/W.

The dissipation due to forward voltage drop and leakage current is given by $P_F + P_i = P_{tot(max)}$. From published data $P_{tot(max)} = 8 \cdot 6$ W. For high-frequency operation at 1 A average current $P_{tot(max)} = 3 \cdot 1$ W; therefore the additional dissipation allowed in the rectifier is $8 \cdot 6 - 3 \cdot 1 = 5 \cdot 5$ W. For a square wave $I_{pk} = 2I_o = 2$ A and $V_R = 100$ V. From Figure 95 for $I_{pk} = 2$ A, $Q_1 = 6 \cdot 9$ μC. Substitution in Equation (59) gives $f_{max} = 10$ kHz. It can be seen that, if V_R is doubled, f_{max} will be halved.

Pulse Operation at High Frequency

The amount of minority carrier charge present in the rectifier depends on the value of the peak current (Figure 95). With pulse operation, where the off–to– on duty cycle ratio is greater than 1 : 1, the same average current I_o as in square-wave operation will give a higher value of peak current. This

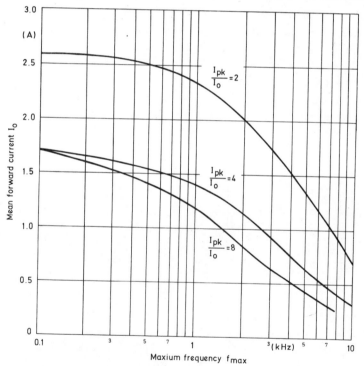

FIGURE 99. Current derating for pulse operation, with reverse voltage V_R of 100 V, maximum ambient temperature of 55 °C, and $R_{th(i)} + R_{th(h)} = 7$ degC/W. For f_{max} at any other reverse voltage V_{R1}, multiply the value read off by $100/V_{R1}$.

means that the rectifier will have to withstand higher dissipation. The derating curves for a BYX38 mounted on a heatsink with a thermal resistance giving $R_{th(h)} + R_{th(i)} = 7$ degC/W are shown in Figure 99 for various values of peak-to-mean current ratio. This thermal resistance is typical of the heatsinks with which the BYX38 would be used.

Rectification Efficiency

The rectification efficiency η_f is defined as

$$\eta_f = \frac{\text{average output voltage at frequency } f}{\text{average output voltage at low frequency } f_{low}}. \tag{63}$$

Figure 100 shows the input and output voltage waveforms of full-wave rectifier operation. When current is transferred from rectifier diode 1 to rectifier diode 2 at the end of the half-cycle, rectifier diode 1 presents a low-impedance path during the time t_2; therefore during this period there is virtually a short-circuit across the whole of the transformer winding, and the current is limited only by the transformer winding impedance. During $t_{rr} - t_2$ the device is recovering and there will be some output voltage. To simplify, assume the device is a short-circuit throughout t_{rr}. This will give a pessimistic result for the rectification efficiency. The current will flow through the load in the normal way as soon as rectifier diode 1 recovers to its blocking state.

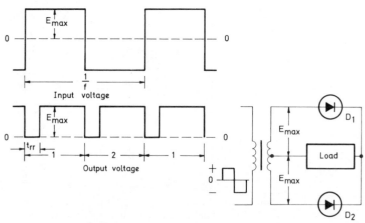

FIGURE 100. Influence of recovery time t_{rr} on rectification efficiency, in the full-wave rectifier circuit shown.

The rectification efficiency is not affected very much at low frequency, as the period of conduction of the rectifier diode is long compared with t_{rr}. However, at high frequencies t_{rr} may be comparable with the period of conduction.

At low frequency

$$\text{average output voltage} \approx E_{max}.$$

At high frequency

$$\text{average output voltage} = \frac{E_{max} \times (1/f) - 2E_{max}t_{rr}}{1/f}$$

$$= E_{max} - 2E_{max}.ft_{rr}.$$

$$\eta_{f} = \frac{E_{max} - 2E_{max}.ft_{rr}}{E_{max}} ; \tag{64}$$

therefore

$$\eta_{f}(\%) = (1 - 2ft_{rr}) \times 100. \tag{65}$$

The above expression indicates that η_{f} is directly dependent on the recovery time t_{rr}. Since t_{rr} varies from one rectifier to another, the rectification efficiency will vary from one diode to another. Figure 101 shows rectification efficiency curves for three samples of BYX38, having t_{rr} values of 25, 17, and 12 μs. These curves apply for $I_R = I_{pk}$. If I_R is greater than I_{pk}, then t_{rr} reduces exponentially, and the rectification efficiency is improved.

Sinusoidal Operation at High Frequencies

In square-wave operation the charge in the device corresponding to any forward current I_{pk} is independent of frequency. In sinusoidal operation the charge in the device varies not only with peak current but also with frequency. It is shown on page 160 that an approximate relationship can be established between the two forms of operation.

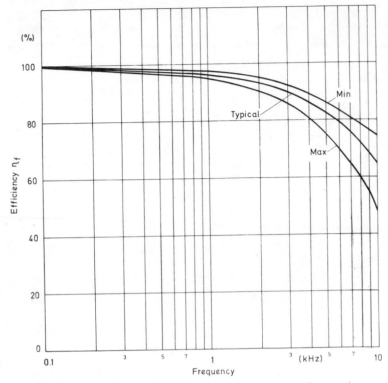

FIGURE 101.   Rectification efficiency as a function of operating frequency for square-wave operation. The curves are for the rectifiers with minimum, typical, and maximum extracted charge (see Figure 91).

Reverse Switching Transient Measurement

The circuit used for the measurement of the minority carrier charge in a rectifier under sinusoidal operation is shown in Figure 102.   Rectifier A is the device under test and rectifier B is used to present a balanced load to the high-power variable-frequency generator.   A fast switching diode C with high back resistance is used to cut off the positive half-cycle of the voltage waveform that appears across R_3, so as not to overload the oscilloscope amplifier when measuring the reverse switching transient.

The extracted charge was measured for the condition where the maximum possible reverse current I_R that can flow through the rectifier is limited to a value equal to the peak forward current I_{pk}.   The waveforms across R_3 are shown in Figure 103, and the shaded area represents the extracted charge.

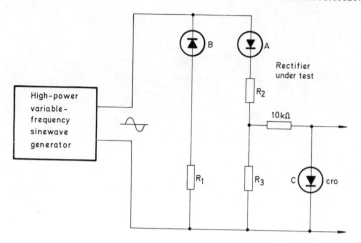

FIGURE 102.   Measurement of minority carrier charge under
sinusoidal conditions.

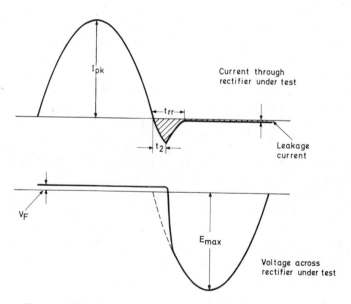

FIGURE 103.   Current and voltage waveforms of rectifier
under test in the circuit of Figure 102.

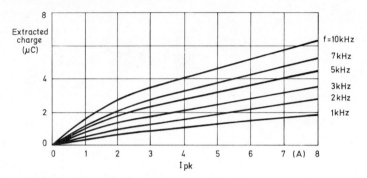

FIGURE 104. Extracted charge as a function of peak current for sine-wave operation at various frequencies.

Experimental Results

Figure 104 shows curves of extracted charge against current I_{pk} for various frequencies.   These curves are plotted for a BYX38 rectifier diode that exhibited the worst high-frequency characteristic.   The charge will vary from rectifier diode to rectifier diode because of the spread in the internal parameters of the diodes.   It is seen that the amount of charge extracted increases with increasing current and frequency.   From these experimental results it was observed that, for any fixed peak current I_{pk}, the extracted charges at frequencies f_1 and f_2 for any particular rectifier diode are related by the following expression:

$$Q_{f_2} \approx Q_{f_1} \left(\frac{f_2}{f_1}\right)^{\frac{1}{2}}.$$ (66)

It was further observed that, if the current at which the charge is measured is halved and the frequency at which the charge is measured is doubled, then in the two cases the extracted charge is approximately the same. Figure 105 shows the transient waveforms for $I_{pk} = 6 \cdot 3$ A, $f = 5$ kHz and also for $I_{pk} = 3 \cdot 14$ A, $f = 10$ kHz.   The shapes of the transients, and the extracted charges, are approximately the same.   The only common factor between the two operations is the rate of current at $t = 0, \pi, 2\pi, \ldots$.
Thus

$$Q_{f_2} \approx Q_{f_1}$$ (67)

when

$$\tfrac{1}{2}I_{pk1} = I_{pk2} \text{ and } 2f_1 = f_2$$

For any value of current I_{pk} and frequency f the rate of change of current at $t = 0$ can be determined by the following equation:

$$\left(\frac{di}{dt}\right)_{t=0} = 2\pi f I_{pk}. \tag{68}$$

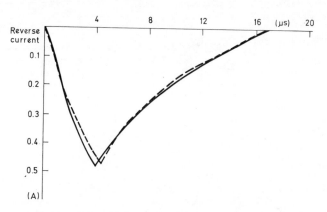

FIGURE 105.   Current transients at 5 kHz, 6.3 A (solid line) and 10 kHz, 3.14 A (dashed line).

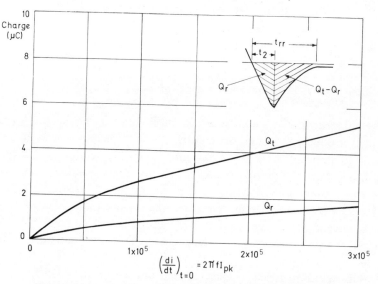

FIGURE 106.   Extracted charge as a function of $(di/dt)^{t+0}$

Figure 106 shows the variation in Q_t and Q_r with $(di/dt)_{t=0}$, and Figure 107 the variation in t_2 and t_{rr} with $(di/dt)_{t=0}$, where Q_t is the total charge extracted during recovery time t_{rr}, and Q_r is the amount of charge extracted during the period t_2 before the diode resistance builds up rapidly to its high value. All four of the above parameters will vary from rectifier diode to rectifier diode.

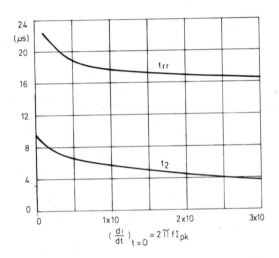

FIGURE 107. t_2 and t_{rr} as functions of $(di/dt)_{t=0}$.

Dissipation Due to Reverse Switching Transients

The minority carriers constitute the mechanism by which the forward current is conducted through the rectifier diode. The concentration of the minority carriers will vary as the current through the rectifier diode varies. Since in sinusoidal operation the current is continuously changing, the charge in the device due to minority carriers is also continuously changing. Therefore less charge is available for extraction when a sinusoidal forward current of peak value I_{pk} is passed through the rectifier diode, compared with the charge available for extraction when a square wave of forward current of value I_{pk} is passed.

The dissipation caused by the transient under sinusoidal operation will therefore be less than for square-wave operation.

Estimation of Dissipation Loss

As in the case of square-wave operation, similar difficulties are encountered in predicting the dissipation with accuracy. The method adopted is, again, one which is found to give a safe answer. For the purpose of estimating the loss the following assumptions are made.

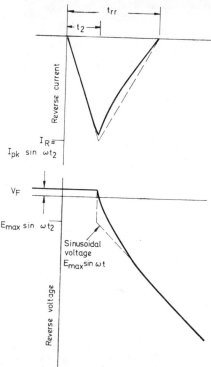

FIGURE 108. Approximation for estimating reverse transient dissipation, for sine-wave operation.

In Figure 108 the diode junction is assumed to be a short-circuit during time t_2. This implies that the dissipation loss during time t_2 is ignored. This is a reasonable assumption, since the dissipation loss during this time is small compared with the total reverse dissipation loss and therefore may be neglected. Further, the reverse voltage is assumed to rise instantaneously to the value of the applied sine-wave voltage at time t_2, and then to follow the applied voltage. The current is assumed to decrease linearly from I_R to zero in a time $t_{rr} - t_2$. The approximations are indicated in Figure 108 by dashed lines.

Suppose that the applied voltage is $v = V_{max}\sin(2\pi f)t$, and the forward current through the rectifier diode is $i = I_{pk}\sin(2\pi f)t$.

The values of t_2 and t_{rr} can therefore be obtained from Figure 107, using the value of $(di/dt)_{t=0}$ determined from Equation (68).

The reverse current I_R at time $t_2 = I_{pk}\sin(2\pi f)t_2$.

Assume a linear decrease during period $t_{rr} - t_2$ and $t_{rr} \geqslant t \geqslant t_2$, then

$$\text{reverse current } i_R = I_{pk}\ \frac{t_{rr} - t}{t_{rr} - t_2}\ \sin(2\pi f)t_2 \tag{69}$$

and

$$\text{reverse voltage } V_R \text{ at time } t = V_{max}\sin(2\pi f)t. \tag{70}$$

The dissipation W_R at frequency f caused by the reverse transient is given by

$$W_R = f\int_{t_2}^{t_{rr}} i_R V_R\,dt.$$

Substituting for i_R and V_R from Equations (69) and (70) gives

$$W_R = \frac{fV_{max}I_{pk}\sin(2\pi f)t_2}{t_{rr} - t_2}\int_{t_2}^{t_{rr}}(t_{rr} - t)\sin(2\pi f)t\,dt\,;$$

hence

$$W_R = fV_{max}I_{pk}\sin(2\pi f)t_2\left\{\frac{\cos(2\pi f)t_2}{2\pi f} - \frac{\sin(2\pi f)t_{rr} - \sin(2\pi f)t_2}{(2\pi f)^2(t_{rr} - t_2)}\right\}. \tag{71}$$

Equation (71) is graphically represented in Figure 109. The values for plotting these curves were obtained by the procedure shown below.

Suppose it is required to determine W_R for a rectifier conducting a sinusoidal forward current of peak value $I_{pk} = 3{\cdot}14$ A at 10 kHz and peak applied voltage of 100 V.

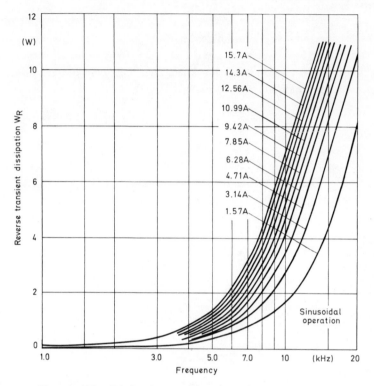

FIGURE 109. Dissipation as a function of operating frequency, for reverse voltage V_R of 100 V, at various peak currents. For dissipation at any other reverse voltage V_{R1}, multiply the value read off by $V_{R1}/100$.

From Equation (68)

$$\left(\frac{\mathrm{d}i}{\mathrm{d}t}\right)_{t=0} = 2\pi10^4 \times 3\cdot14 = 1\cdot975 \times 10^5.$$

From Figure 107 for $\left(\dfrac{\mathrm{d}i}{\mathrm{d}t}\right)_{t=0} = 1\cdot975 \times 10^5$, $t_{rr} = 17\ \mu s$, and $t_2 = 4\cdot4\ \mu s$

$$2\pi f = 6\cdot28 \times 10^4,$$
$$\sin(2\pi f)t_2 = 0\cdot272,$$
$$\cos(2\pi f)t_2 = 0\cdot962,$$
$$\sin(\pi f)t_{rr} = 0\cdot876,$$
$$(2\pi f)^2 = 39\cdot4 \times 10^8,$$
$$t_{rr} - t_2 = 12\cdot6 \times 10^{-6}.$$

Substituting the above values in Equation (71) gives $W_R = 2 \cdot 8$ W. The dissipation loss W_R calculated from Equation (71) is generally higher than the actual loss. The difference between the estimated loss and the actual loss will depend on how close the actual recovery voltage approached the assumed path of voltage recovery.

Derating of Rectifier Diodes

The dissipation loss W_R estimated in the previous section must be taken into account when operating rectifiers at frequencies at which this loss is significant. Figure 110 shows the derating curves for a sine wave operation up to an ambient temperature of 55 °C. The procedure adopted for deriving these curves is illustrated below by an example.

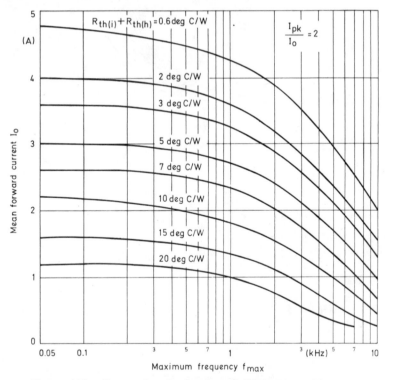

FIGURE 110. Current derating for sine-wave operation, with reverse voltage V_R of 100 V, maximum ambient temperature of 55 °C, and various values of $R_{th(i)} + R_{th(h)}$. For f_{max} at any other reverse voltage V_{R1}, multiply the value read off by $100/V_{R1}$.

Suppose that a BYX38 rectifier diode is mounted on a heatsink with a thermal resistance $R_{th(h)}$ of 4·4 degC/W, giving $R_{th(h)} + R_{th(i)} = 5$ degC/W. It is required to determine the maximum frequency at which a mean current of 2 A can be safely passed through the rectifier diode.   The supply voltage is 100 V peak, and the maximum ambient temperature is 55 °C.

From the published data, for an ambient temperature of 55 °C, and low frequency operation

$$R_{th(h)} + R_{th(i)} = 5 \text{ degC/W}.$$

Dissipation $P_{tot(max)}$ due to forward voltage drop and leakage current $= 8·6$ W.   At average current $I_o = 2$ A,

$$P_{tot(max)} = 5·5 \text{ W};$$

therefore

additional dissipation permissible $= 8·6 - 5·5 = 3·1$ W.

For single-phase sinusoidal operation

$$\frac{\text{peak current per rectifier } I_{pk}}{\text{average current per rectifier } I_o} = 3·14;$$

therefore

$$I_{pk} = 3·14 \times 2 = 6·28 \text{ A}.$$

From Figure 109 for $W_R = 3·1$ W and $I_{pk} = 6·28$ A

$$f_{max} = 8·8 \text{ kHz}.$$

Rectification Efficiency

The rectification efficiency η_f is dependent on the reverse recovery time t_{rr} as in the case of square-wave operation.   Equation (65), derived for the efficiency under square-wave operation, applies equally to sine-wave

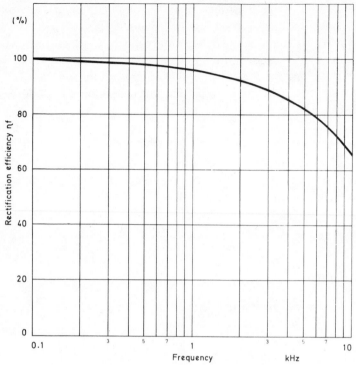

FIGURE 111.   Rectification efficiency as a function of operating
frequency, for sine-wave operation.

operation.   Figure 111 shows the rectification efficiency for a full-wave
rectifier circuit operated with sinusoidal voltage.   This is for a device
with $t_{rr} = 17$ μs; the same device gave $t_{rr} = 25$ μs under square-wave
operation.   The efficiency curve shown will vary slightly with current,
since, at any given frequency, t_{rr} will tend to decrease with increasing peak
current (Figure 107).

The rectification efficiency for sinusoidal operation is higher than that
for square-wave operation.   This is because $t_{rr(sine\ wave)}$ is less than
$t_{rr\ (square\ wave)}$.

Relationship Between Charge Extracted in Square-wave and Sine-wave Operation

The concentration of the minority carriers in the rectifier diode at any
instant is largely dependent on the current flowing through the diode at

that instant. In square-wave operation, when the rectifier diode is switched to the forward conducting state, the concentration of minority carriers reaches a steady state once the forward current is established. In sine-wave operation the amplitude of the forward current is continuously changing; therefore the concentration of the minority carriers will tend to vary continuously. The concentration increases as the forward current increases to its peak value, and then tends to decrease as the current reduces to zero. The charge in the device during this period can reduce only at the rate allowed by the decay time constant τ.

It is shown below that, if both the extracted charge–peak current relationship of a rectifier diode under square-wave operation (Figure 95) and the time constant (Figure 93), are known then the charge extracted for a sine wave of current with a peak value I_{pk} and frequency f can be estimated.

Estimation of Charge

Consider a rectifier diode which is passing a forward current that varies sinusoidally. Since the rate of change of current in the region of the peak current is small, it is reasonable to assume that the charge corresponding to the peak current is established in the rectifier at $\omega t = \pi/2$. It therefore follows that any excess charge present in the device, when the current falls to zero, occurs during the period $\omega t = \pi/2$ to $\omega t = \pi$.

Figure 112 is drawn for the period $\pi/2$ to π and shows the charge required to maintain current during the second half of the conducting-half-cycle of the sine wave. For a sine wave of current of peak value I_{pk} and frequency f, the sine wave can be regarded as a number of small steps each of duration τ. At each step, the charge Q_m required to maintain current can be plotted against time as shown in Figure 112, using the value of $Q_m = Q_i$ obtained from Figure 95.

In the interval τ just before the current reaches zero, a charge Q_i is required to maintain current. This charge will reduce to $Q_1 e^{-1}$ when the current passes through zero. In the next interval τ, the excess charge is $Q_2 - Q_1$, and this charge reduces to $(Q_2 - Q_1)e^{-2}$ when the current passes through zero. The charge Q_e extracted from the rectifier under sinusoidal operation can therefore be expressed approximately by the following series:

$$Q_e \approx Q_1 e^{-1} + (Q^2 - Q_1)e^{-2} + (Q_3 - Q_2)e^{-3} + (Q_4 - Q_3)e^{-4} + (Q_5 - Q_4)e^{-5}.$$
$$(72)$$

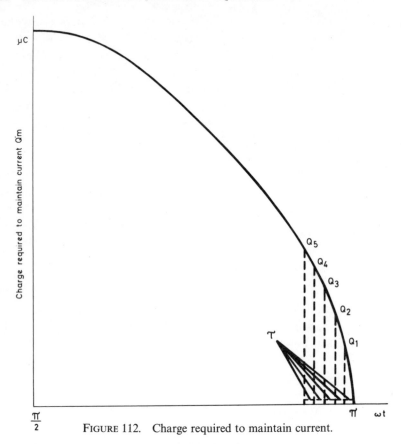

FIGURE 112.   Charge required to maintain current.

In deriving the above expression it is assumed that the charge which is recombined during the reverse recovery period t_{rr} under sinusoidal operation is the same as that under square-wave operation.

Comparison of Estimated and Experimental Results

The estimated and experimental results are given in Table 6.   There is fair agreement between the results.   The examples below illustrate the method of deriving these figures.

It is required to estimate the charge extracted under sinusoidal operation at (a) $f = 5$ kHz, $I_{pk} = 5$ A, $\tau = 16$ μs, and (b) $f = 10$ kHz, $I_{pk} = 5$ A, $\tau = 16$ μs.

Table 6.   Comparison of estimated charge Q_{est} and measured charge Q_{exp}

I_{pk} (A)	$f = 10$ kHz		$f = 7$ kHz		$f = 5$ kHz		$f = 3$ kHz		$f = 2$ kHz		$f = 1$ kHz	
	Q_{est}	Q_{exp}	Q_{est}	Q_{exp}	Q_{est}	Q_{exp}	Q_{est}	Q_{exp}	Q_{est}	Q_{exp}	Q_{est}	Q_{exp}
0·5	1·11	0·90	0·96	0·70	0·84	0·40	0·59	0·42	0·46	0·30	0·26	0·26
1·0	1·64	1·60	1·49	1·22	1·30	1·00	1·01	0·80	0·78	0·60	0·47	0·45
2·0	2·45	2·60	2·20	2·10	1·96	1·70	1·55	1·30	1·26	1·05	0·79	0·77
3·0	3·15	3·40	2·81	2·80	2·45	2·30	1·98	1·72	1·59	1·35	1·05	0·95
4·0	3·85	4·00	3·38	3·35	2·92	2·75	2·34	2·10	1·87	1·60	1·27	1·12
5·0	4·56	4·52	3·96	3·80	3·36	3·15	2·68	2·40	2·14	1·90	1·43	1·30
6·0	5·32	5·10	4·59	4·25	3·77	3·60	2·99	2·75	2·39	2·15	1·62	1·48
7·0	6·06	5·60	5·15	4·70	4·39	4·00	3·32	3·10	2·65	2·40	1·78	1·62

The values are in microcoulombs.

Example (a).   The equation of the current for the period $\omega t = \pi/2$ to π is

$$i = 5\cos(2\pi 5 \times 10^3)t.$$

The quarter-period at 5kHz is 50 μs.

$$i_1 = 5\cos\{\pi \times 10^4(50 - 16) \times 10^{-6}\} = 2\cdot395 \text{ A};$$

therefore

Q_1 (from Figure 95) $= 7\cdot7\ \mu$C,
$i_2 = 5\cos\{\pi \times 10^4(50 - 32) \times 10^{-6}\} = 4\cdot22$ A,
$Q_2 = 11\cdot5\ \mu$C,
$i_3 = 5\cos\{\pi \times 10^4(50 - 48) \times 10^{-6}\} = 4\cdot99 \approx 5$ A,
$Q_3 = 13\cdot1\ \mu$C.

From Equation (72)

$$Q_e \approx 7\cdot7e^{-1} + (11\cdot5 - 7\cdot7)e^{-2} + (13\cdot1 - 11\cdot5)e^{-3} \approx 3\cdot3\ \mu\text{C}.$$

The experimental value at 5 kHz, 5 A, $Q_e = 3\cdot15\ \mu$C.

Example (b).   The equation of current for the period $\omega t = \pi/2$ to π is $i = 5\cos(2\pi 10^4)t$.   The quarter-period at 10 kHz is 25 μs.

$$i_1 = 5\cos\{2\pi 10^4(25 - 16) \times 10^{-6}\} = 4\cdot22 \text{ A},$$
$$Q_1 = 11\cdot5 \ \mu\text{C}.$$

i_2 must be taken as $I_{pk} = 5$ A, since $2\tau > 25 \ \mu$s; therefore $Q_2 = 13\cdot1 \ \mu$C. From Equation (72)

$$Q_e = 11\cdot5e^{-1} + (13\cdot1 - 11\cdot5)e^{-25/16} \sim 4\cdot56 \ \mu\text{C}.$$

The experimental value at 10 kHz, 5 A is $4\cdot52 \ \mu$C.

Diode and Thyristor in Series

Knowledge of the reverse recovery characterics of rectifier diodes is also of particular interest in thyristor applications (not necessarily high-frequency ones).   Where a rectifier diode is connected in series with a thyristor to protect the thyristor against reverse breakdown, the extra components to be used depend on the voltages present.

The first case to be considered is where the applied repetitive peak reverse voltage V_{RR} is greater than the repetitive peak reverse voltage rating of the thyristor, but less than that of the diode.   The rectifier diode must then recover before the voltage across the thyristor has risen to the thyristor's repetitive peak reverse voltage rating.

It has been found that the recovery time of thyristors is generally shorter than that of the higher-voltage rectifier diodes.   This voltage condition can therefore be met by connecting a small capacitor across the thyristor to limit the voltage across the thyristor, but no capacitor across the rectifier diode.   The value of the capacitor is given by

$$C \geqslant \frac{Q}{V_{RR}}, \tag{73}$$

where Q is the stored charge of diode, corresponding to the appropriate forward current.   For design purposes, this must be the maximum value for the type of diode used.

The leakage current of the diode is usually less than that of the thyristor, so that most of the crest working voltage appears across the diode.   No shar-

ing resistors are needed if the applied V_{R_w} is less than the diode V_{R_w} rating.
The second case is where the applied repetitive peak reverse voltage is
greater than the repetitive peak reverse voltage rating of the thyristor and
greater than that of the diode.   Capacitors and resistors are thus required
across both the thyristor and the diode, as described on page 166.

Summary

From this section it can be concluded that the performance of rectifier
diodes at higher frequencies is limited by the dissipation caused by reverse
switching transients.   This dissipation is greater for pulse and square-wave
operation than for sine-wave operation.

The rectification efficiency of the rectifier diodes is directly dependent
on the reverse recovery time t_{rr}.   Therefore for satisfactory operation at
high frequencies both t_{rr} and the extracted charge should be kept to a
minimum.   Then both the rectification efficiency and the maximum allowable frequency limit will show improvement.

The reverse dissipation under specified conditions varies from diode
to diode of any one type (Refs. 108 to 110).   It is thus necessary to know
the reverse dissipation and maximum frequency limit for the worst rectifier
diode (from the high-frequency performance point of view) within the
specification of any particular type of rectifier diode.

General Notes on the use of Rectifier Diodes

In this section the operation of silicon rectifier diodes is considered
when the output voltage requirement exceeds the voltage rating of a single
rectifier diode or when the output current requirement is higher than the
current rating of a single device (Refs. 2, 100).   General notes on rectifier
diode protection against circuit failures and transient voltages are also
included.

Series Operation

Rectifier diodes are connected in series when the output voltage requirement exceeds the voltage rating of a single rectifier diode.   In this case
it is necessary to connect two or more rectifier diodes in series.

There is an inevitable spread in the characteristics of any type of rectifier diode and it is almost certain that, when two or more diodes are connected together, their forward, reverse, or both characteristics will differ from one another. In series operation differences in the forward characteristic are of no consequence since each diode carries a common current.

In the reverse direction, however, the voltage across each diode depends on the leakage current, and, since there is a spread in the leakage current rating of the rectifier diodes, it is necessary to connect a resistor across each diode to ensure that the rated crest working voltage is not exceeded. This does not apply to the avalanche diodes.

In addition, if the rectifier diode is reverse-biased immediately after it has been carrying forward current, it requires a finite time to recover to its blocking state. This time will depend on the stored charge, which varies from one diode to another. Thus, if a reverse voltage is applied across several rectifier diodes in series, then the diode with the least stored charge will recover first and will be subjected to the full applied voltage. This may damage one or more diodes and therefore a capacitor is sometimes connected across each rectifier diode for protection against fast transient voltages.

When only two diodes are connected in series, the capacitors are not always needed, because the instantaneous value of the applied voltage, immediately after blocking, frequently does not exceed the voltage rating of one diode.

Determination of Reverse Voltage-sharing Resistor

To determine the value of the reverse voltage-sharing resistor a chain of n rectifier diodes connected in series is shown in Figure 113(a). Across each diode is connected a resistor R with a tolerance $\pm \Delta R$.

The diode D_1 is considered to be an ideal rectifier (that is with no leakage current). The remaining diodes are considered to have maximum leakage current $I_{R(max)}$ at maximum junction temperature. The maximum voltage V_1, to which diode D_1 is subjected, will occur when the resistor across the diode D_1 is $R + \Delta R$ and that across each of the other diodes is $R - \Delta R$. This condition is shown in Figure 113(b), where the voltage across diode D_1 is

$$V_1 = (I_1 + I_{R(max)})(R + \Delta R) \tag{74}$$

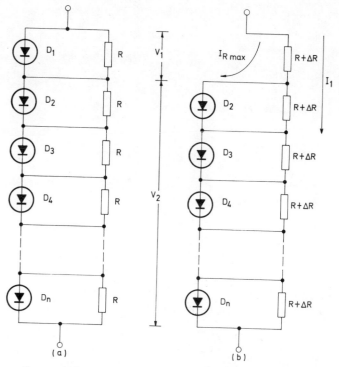

FIGURE 113.   Determination of voltage-sharing resistor.

and the voltage across the rest of the chain is

$$V_2 = I_1 (R - \Delta R)(n - 1);\qquad\qquad(75)$$

therefore

$$I_1 = \frac{V_2}{(R - \Delta R)(n - 1)}.\qquad\qquad(76)$$

Substitution of (76) in (74) gives

$$V_1 = \left\{\frac{V_2}{(R - \Delta R)(n - 1)} + I_{R(max)}\right\}(R + \Delta R)\qquad(77)$$

and multiplying (77) by $R/(R + \Delta R)$ gives

$$\frac{V_1 R}{R + \Delta R} = \frac{V_2 R}{(R - \Delta R)(n - 1)} + I_{R(max)}R. \tag{78}$$

If $\beta = \Delta R/R$, Equation (78) then becomes

$$\frac{V_1}{1 + \beta} = \frac{V_2}{(1 + \beta)(n - 1)} + I_{R(max)}R. \tag{79}$$

Since $V_1 + V_2 = $ applied reverse voltage V_R

$$R = \frac{1}{I_{R(max)}}\left\{\frac{V_1}{1 + \beta} - \frac{V_R - V_1}{(1 - \beta)(n - 1)}\right\}. \tag{80}$$

The maximum value of V_1 that is permissible is the crest working voltage, V_{RW} rating of the rectifier diode; therefore

$$R \leqslant \frac{1}{I_{R(max)}}\left\{\frac{V_{RW}}{1 + \beta} - \frac{V_R - V_{RW}}{(1 - \beta)(n - 1)}\right\}. \tag{81}$$

For this equation to yield practical values of R,

$$\frac{V_{RW}}{1 + \beta} > \frac{V_R - R_{RW}}{(1 - \beta)(n - 1)}$$

or

$$n - 1 > \frac{V_R - V_{RW}}{V_{RW}}\frac{1 + \beta}{1 - \beta}.$$

therefore

$$n > 1 + \frac{V_R - V_{RW}}{V_{RW}}\frac{1 + \beta}{1 - \beta} \tag{82}$$

Equation (82) indicates the minimum possible number of diodes which have to be connected in series for a given resistor tolerance β.

Determination of Capacitor for Transient Voltage Sharing

Figure 144(a) shows rectifier diodes connected in a series chain. Across each diode a capacitor C is connected to protect the diodes against voltage transients.

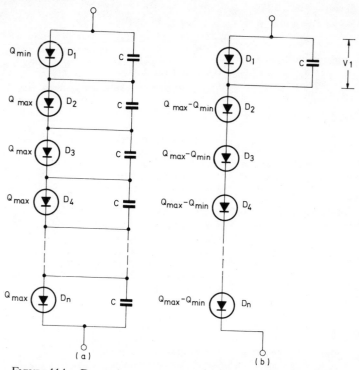

FIGURE 114. Determination of transient voltage-sharing capacitor.

The worst condition will occur when rectifier diode 1 has minimum stored charge Q_{min}, and the rest have maximum stored charge Q_{max}. When a reverse voltage is applied to the chain immediately after forward conduction, the diode with the least stored charge will recover first and will be subjected to some voltage V_1. This voltage V_1 must not exceed the maximum repetitive peak inverse voltage rating V_{RR} of the rectifier diode. This condition is shown in Figure 114(b).

Now,

$$Q_{max} - Q_{min} = CV_1$$

and, since $V_1 \leqslant V_{RR}$,

$$C \geqslant \frac{Q_{max} - Q_{min}}{V_{RR}}. \tag{83}$$

In deriving the above expression it has been assumed that the recombination time of the rectifier diodes is about the same.

When the rest of the rectifier diodes have recovered, the transient voltage will be shared equally, provided that the value of C determined in Equation (83) is at least ten times the junction capacitance. Usually, the value of C given by Equation (83) is very large compared with the junction capacitance.

Example. Determine the number of BYX38 rectifier diodes required in series, and the value of R to be connected across each diode, for three-phase full-wave operation in which each arm of the bridge is subjected to a peak reverse voltage V_R of 3600 V.

For BYX38 rectifier diodes, use $V_{RW} = 400$ V and $I_{R(max)} = 600$ μA. Let the tolerance on the resistor R be $\pm 5\%$. From Equation (82)

$$n > 1 + \frac{3600 - 400}{400} \frac{1 + 0.05}{1 - 0.05} > 9.88$$

Let $n = 10$. Then, from Equation (81),

$$R \leqslant \frac{1}{600 \times 10^{-6}} \left\{ \frac{400}{1 \times 0.05} - \frac{3600 - 400}{(1 - 0.05)9} \right\} \leqslant 10 \text{ k}\Omega.$$

The power dissipated P_R in the resistor R in a three-phase full-wave bridge is given by

$$P_R \approx \frac{0.402 V_{RW}^2}{R}$$

$$\approx \frac{0.402(400)^2}{10 \times 10^3} \approx 0.643 \text{ W}.$$

If this dissipation is not tolerable, n must be increased. Let $n = 11$; then

$$R \leqslant \frac{1}{600 \times 10^{-6}} \left\{ \frac{400}{1 + 0.05} - \frac{3600 - 400}{(1 - 0.05)10} \right\} \leqslant 73 \cdot 3 \text{ k}\Omega.$$

Choose $R = 68$ kΩ $\pm$ 5%. The power dissipated in R is then

$$P_R \approx \frac{0 \cdot 402(400)^2}{71 \cdot 4 \times 10^3} \approx 0 \cdot 09 \text{ W}.$$

Thus, series connection of eleven BYX38 rectifier diodes in each arm of the bridge, with 68 kΩ $\pm$ 5% across each rectifier diode, gives a suitable solution.

Parallel Operation

Rectifier diodes are connected in parallel to allow higher power to be handled or to increase the reliability of installation. The differing characteristics may lead to unequal current sharing and to possible over-running and destruction of one or more diodes.

There are a number of methods of achieving a more equal sharing of current. Some of the methods are discussed and practical examples are given.

Spread in Forward Characteristics

It is the spread in the forward characteristic that is of greatest importance in parallel diode operation. Unequal current sharing can lead to a marked increase in the temperature of diodes that are carrying disproportionately large shares of the total current, and the temperature change will accentuate the characteristic differences. A cumulative process of current and temperature increase may continue to a level at which the maximum current and temperature ratings of the device are exceeded.

It is worth noting that the maximum operating conditions given in rectifier diode data are determined by the 'worst' diode (that is the one with the greatest forward voltage drop V_f at current i), since this diode will have the maximum dissipation.

All 'better' diodes would reach maximum dissipation only at currents above the maximum current limit.

When the two diodes are connected in parallel, it is possible (though not probable) that they will be 'best' and 'worst' diodes, that is diodes with the lowest and highest values of the forward voltage V_f respectively. However, this least favourable combination of diodes will be considered. In most practical cases, two diodes with a smaller difference in characteristics will be used, and the performance will be better than that obtained with opposite limit samples.

If for two diodes in parallel D_1 is the best diode and D_2 the worst, with a current i_1 flowing through the best diode and i_2 flowing through the worst, and if the average total output current is I_{av}, then

$$I_{av} = i_{av(1)} + i_{av(2)}$$

and

$$i_{av(1)} > i_{av(2)}.$$

It follows that, if I_{av} is twice the rated maximum current for a single diode, then the current $i_{av(1)}$ through the best diode will be in excess of the maximum rating, while $i_{av(2)}$ will be correspondingly within the rating. The junction temperature $T_{j(1)}$ of D_1 will thus be higher than $T_{j(2)}$, and the forward characteristic of D_1 will undergo a greater shift than that of D_2.

The series of effects rapidly leads to a state in which the maximum divergences of i_1 and i_2, and $T_{j(1)}$ and $T_{j(2)}$, are established. This is usually a stable state for diodes well below the maximum rated current, but, if the operation is close to the rated values and if the initial differences in the characteristics are great enough, thermal runaway may take place and the diodes may be successively destroyed—the better diode first, and then (when it alone carries the full load current) the worse diode.

Methods of evaluation of current difference are given in Ref. 100. These were based and verified by measurements. It should be noted, however, that the accuracy of the calculations which are based on measurements depend on the accuracy with which voltage and current are measured. Therefore, in most cases, of great interest are the results and the suggested measures to keep both current and temperature within the device ratings.

Because large differences in current can occur between two or more rectifier diodes connected in parallel, steps must be taken to keep both current and temperature within the device ratings.

One remedial measure is the reduction of temperature difference by the sharing of a common heatsink. It was shown in Ref. 100 that a much greater current difference occurs when the diodes are mounted on separate heatsinks, and that the difference is much reduced when the diodes are on one heatsink and when the spacing between them is kept to a minimum.   However, this method is only one of several possibilities.   Six methods, including heatsink sharing ('thermal coupling') will now be discussed and assessed. The methods are as follows: (i)   selection of diodes for minimum spread of characteristics; (ii)   temperature derating, by use of larger heatsinks or increased air cooling; (iii) thermal coupling; (iv) use of series resistors; (v) use of balancing transformers; (vi) current derating, in which the total current is shared between an increased number of parallel diodes.

In practice, it may be desirable to combine two or more of these methods.

Selection of Diodes

It is possible to limit the current difference by using diodes that have only a small difference between their forward characteristics.   Thus diode combinations that have a forward voltage difference at 100 A of less than, say, 70 mV have small average current ratios.   The use of selection, therefore, has the advantage that the diodes can be operated with little or no current derating.   In the ideal case diodes with *no* differences between their forward characteristics would be used, and neither derating nor any other measure would be required.

The production of one type of diode could be divided into two or three subtypes, according to a fixed difference in the forward voltage, and, in any one rectifier installation, diodes from only one of these subtypes would be used.

Selection, however, is troublesome to the rectifier user.   To the rectifier diode manufacturer, it presents economic and production problems of some magnitude.   Thus in place of X types of rectifier diode, he would have to provide $3X$ or more types.   And, in any event, there are unavoidable shifts in the mean value of any characteristic during an extended production run, so that it would sometimes be impossible to supply diodes of the subtypes that lie at the ends of the total spread of the original diode type.

For these reasons, it is not generally desirable to resort to selection.

Thermal Coupling

It has already been said that the sharing of heatsinks and the reduction of the distance between diode centres has a favourable effect.   The mounting base temperatures will be more or less equalised, and the junction temperature of the best diode will be kept down.   The reduced difference in junction temperatures will counteract the divergence of forward characteristics, so that current differences, as already mentioned, will be considerably reduced.

The value of $R_{th(mb-amb)}$ for a single heatsink carrying n diodes must be n th fraction of the nominal $R_{th(mb-amb)}$ required for a single diode.

Thermal coupling, in combination with other methods of protection, is strongly recommended.   Nevertheless, care must be taken in the calculation of the heatsink (Ref. 11) to ensure that the diode spacing is not so small that the efficiency of the heatsink is seriously impaired.

Series Resistance

Circuit differences can be reduced by the connection of equal resistances in series with each diode.   The differing internal resistances of the diodes will then form smaller parts of the total resistance of each path, and the currents will therefore show less divergence.   If the added resistance is several times the internal resistance of the diode, then substantial equalisation of currents can be achieved, at the cost, however, of increased losses.

The determination of the value of series resistance is based on two principles: (i) the dissipation must be equal in each diode, and (ii) the current through the best diode must not exceed the maximum published rating. For equal dissipation the calculations required are extensive, but, for practical purposes, satisfactory equalisation of diode temperatures is obtained if the value of R is chosen so that, at the maximum rated peak forward current, the peak voltage drop across the resistor is 1·5 V.   This procedure is adequate for all types of circuit including three-phase and six-phase configurations.

For diode current equal to the rated values the derating factor is given in terms of the peak current for a given series resistance R.   When each diode is protected by a fuse with a minimum rating of $1·15 \times i_{rms(nom)}$, I_{pk} calculation is based on a series resistance of about 1·5 mΩ.   A derating factor of $d = 0·85 + 0·15/n$ is then used, where n is the number of diodes.

Balancing Transformers

The principle of the balancing transformer is shown in Figure 115.

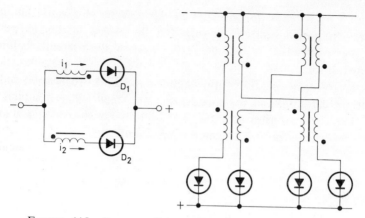

FIGURE 115. Two parallel rectifier diodes with balancing
transformer.

FIGURE 116. Four parallel rectifier diodes with balancing
transformer in closed chain.

Each of the two parallel diodes has in series with it a transformer winding, both windings being on the same core.   If D_1 is the best diode and D_2 the worst, then $V_{o(1)} > V_{o(2)}$, and D_1 will conduct earlier in the positive half-cycle than D_2.   The current i_1, flowing through the winding associated with D_1, will induce a current in the other winding, thus forcing D_2 to conduct.   The current i_2 will set up an opposing field, and i_1 will be reduced. This interaction will tend to correct the unbalance of the diode currents.

For the calculation of the inductance of each winding, the following empirical formula may be used:

$$L = \frac{1}{\omega i_{pk(nom)}\delta} ,\qquad (84)$$

where δ is

$$\frac{i_1 - i_2}{\frac{1}{2}(i_1 + i_2)} .$$

If $\delta \leqslant 0.05$ and $f = 50$ Hz, then

$$L = \frac{63.5}{i_{pk(nom)}} \text{ mH} \qquad (85)$$

and no further derating is required.

When more than two diodes are connected in parallel, a 'closed chain' arrangement is recommended, as shown in Figure 116.   If, in this circuit, any one of n diodes is disconnected by the blowing of a fuse, the circuit continues to provide current sharing, but the remaining diodes will, of course, be overloaded by the $(n-1)$th part of the current of the disconnected diode.

The use of balancing transformers allows the diodes to be used without any current derating, and the added losses are small. The disadvantages are the cost of the transformers, the increase in the bulk of the installation, and the switching transients that are produced.

Current Derating

It is unlikely that, in any installation, one 'best' diode will be connected in parallel with one or several 'worst' diodes. However, it is possible that diodes with *less* than the maximum characteristic difference can run with unacceptably large current differences, and the best diode of those used in an installation is likely to be overrun.

Current sharing can be improved by the techniques that have already been discussed, but, unless a suitable balancing transformer or a sufficiently large series resistor is used, some measure of current derating must be employed to ensure that the best diode of the group operates at the maximum current rating, and all the other diodes at less than the maximum rating.

Two ways of achieving this derating are possible; either the total current that is taken from the installation must be reduced, or the current between a larger number of parallel diodes must be shared. At the design stage of the installation, of course, these two approaches amount to the same thing: the installation must be designed so that the total current and the number of parallel diodes are such that the best diode in the worst probable combination of diodes will run at the maximum current rating of the type.

It can be shown that the diode combinations with the largest differences in their forward characteristics have the most severe derating factors. In the worst cases, it is necessary to reduce the current to about 60% of its original value. These deratings may be calculated in the following way.

Let the measured ratio $i_{1(av)}/i_{2(av)} = y$ for a particular diode pair. If BYZ14 is taken as an example, the maximum average current rating of each diode is 40 A, so that the average total current that may be conducted is $2 \times 40 = 80$ A. The total current that may be conducted in practice is $40 (1 + 1/y)$ A; therefore the derating factor is

$$d = \frac{40\,(1 + 1/y)}{80}.$$

As an example, the required derating for the pair of diodes under half-wave conditions, is given by

$$y = \frac{i_{1\,(av)}}{i_{2\,(av)}} = 2\cdot64;$$

then

$$I_{1\,(av)} = 40 \text{ A},$$

$$I_{2\,(av)} = \frac{I_{1\,(av)}}{y} = \frac{40}{2\cdot64} = 15\cdot1 \text{ A},$$

$$I_{total} = I_{1\,(av)} + I_{2\,(av)} = 40 + 15\cdot1 = 55\cdot1 \text{ A}.$$

For this total current to flow, the nominal rating of 80 A must be derated by

$$d = \frac{55\cdot1}{80} = 0\cdot69.$$

Figure 117 (based on large number of samples of the BYZ14 diode) shows the current derating factor d as a function of the number of diodes n connected in parallel, with, as parameter, the chance of overloading the best diode in any combination. The two dashed lines are the curves for chances of $0\cdot1\%$ and $0\cdot01\%$. The solid line represents a derating of $d = 0\cdot8 + 0\cdot2/n$, which is recommended.

When the recommended derating is applied, then, in the worst case (a 'best' diode paralleled with n-1 'worst' diodes) the best diode will have a maximum dissipation overload as shown in Table 7 for various values of n.

Dissipation overloads of these magnitudes will give thermal runaway only if the diode has the maximum published value of $R_{th(j-c)}$, which again is unlikely. In any event, thermal stability will be substantially improved if the recommendation to use a shared heatsink is followed.

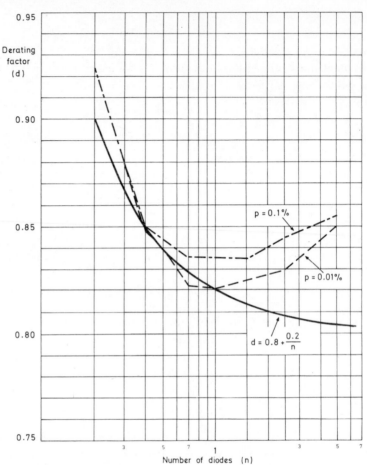

FIGURE 117.   Derating factors (d) for two to sixty diodes in parallel for 0.1% and 0.01% change (p) of overloading the best diode.

Table 7.   Overload in worst case.

Number(n) in parallel	Dissipation overload of best diode ($\%$)
5	0
10	4
50	7·5

This form of diode protection—by current derating—has the advantage that it does not need supplementing by any other protection method (but thermal coupling should also be used to give additional protection). However, there is the obvious disadvantage that more diodes (and therefore a larger installation) will be necessary for any particular required total current.

Surge Characteristics

Equalisation of diode operation has been discussed in terms of average currents. However, diodes with differing forward characteristics will also share surge currents unequally. Similar considerations apply to surge currents as to average currents.

Published surge current ratings are based on the characteristics of the worst diode, since this is the diode that will first reach the maximum dissipation. A better diode will reach this limit at higher currents.

Measurements have shown that current sharing at these high surge currents is somewhat better than at average currents. Satisfactory protection will be given if the fuse in series with each diode is rated at $1 \cdot 15 \times i_{rms(nom)}$.

Recommendations

Which method, or combination of methods, is used is determined mainly by cost and by the required reliability of the installation. The following guidance is offered.

(1) If diodes are connected in parallel without any additional circuit elements, then either of the following deratings should be used: (i) average current to $0 \cdot 8 + 0 \cdot 2/n$ of the nominal published value or (ii) thermal resistance of heatsink to 75% of the nominal required value.

(2) Thermal coupling of diodes is strongly recommended, whatever other measures are employed. With n diodes, the thermal resistance of the shared heatsink should be the nth fraction of the value required for one diode.

(3) A fuse in series with a diode acts as a small series resistance, and improves current sharing. Current derating to $0 \cdot 85 + 0 \cdot 15/n$ is then sufficient.

(4) No derating is necessary when using either (i) a balancing transformer with $L = 20/(\omega i_{pk(nom)})$ H, or (ii) a $\pm 5\%$ series resistor, for

each diode, such that the peak voltage drop across the resistor is 1·5 V at nominal peak current.

(5) The electrical and mechanical layout must be as symmetrical as possible.

(6) Selection of diodes may be resorted to, but, in general, it is more convenient and economic to use other methods.

Rectifier Diode Protection

Protection of the rectifier circuit must be carefully designed, as in certain circuits failure of one device may lead to the destruction of the rest of the rectifier diodes.

To prevent damage to rectifier diodes in the event of short-circuit, they must be protected by fuses. The fuse must blow before the surge current rating of the rectifier diode is exceeded; therefore the correct fuse must be selected, due attention being paid to the surge current rating of the rectifier diode (Refs. 101 to 105).

The rectifier diodes are equally likely to be damaged by the voltage surges occurring in the supply or generated within the equipment itself. Voltage transients are likely to be at a maximum when the equipment is under no-load condition or has a choke input filter. The common sources of voltage transients are listed; (1) interruption of transformer magnetising currents; (ii) energising of transformer primary; (iii) inductive loads in parallel with the rectifier diode input; (iv) load switching on d.c. side of the rectifier diode; (v) energising of a step-down transformer; (vi) hole storage recovery phenomenon; (vii) opening of individual fuse for parallel devices.

The majority of the voltage transients can be reduced by the application of suitable capacitances, or resistances and capacitances in series (Refs. 101 to 105). The position of these components depends largely on the type of transient to be suppressed.

Recent advances in the semiconductor field have led to the development of controlled avalanche rectifier diodes, which have a reverse characteristic very similar to the characteristics of the Zener diode, but occurring at very much higher voltages. This characteristic enables the rectifier diode to absorb a limited amount of transient energy.

The avalanche characteristic will be particularly useful for series operation of rectifier diodes, and will lead to improvements in the transient and steady-state voltage—sharing properties of the rectifier diodes when connected in series.

However, it appears almost certain that both conventional rectifier diodes and the controlled avalanche types will be needed, since the power absorption capability of the controlled avalanche rectifier diode is limited, and is not adequate to absorb all the energy contained in voltage surges occurring on normal low-impedance mains supplies.

Summary

In this section an attempt has been made to illustrate the great potentiality of silicon power rectifier diodes.

For single-phase rectification, the choke input filter is more suitable than the simple capacitor filter, especially where the voltage regulation is important. However, up to a certain power level it is feasible to use the simple capacitor filter, particularly where a d.c. supply of one specified current is required, and where the weight of the equipment has to be kept to a minimum. The limiting factor is the maximum ripple current that the capacitor can handle.

Three-phase rectifier circuits can be effectively used for delivering large amounts of power, provided that adequate protection of the rectifier diodes is included in the circuit. The increased reliability, low running costs, and relatively low cost of the silicon rectifier diodes have made it an economical proposition to consider them for heavy-current applications.

Transistors Used as Rectifier Diodes

Transistors as General-purpose Rectifiers

All transistors, as mentioned previously, consist of two junction diodes, the collector–base diode and the emitter–base diode. Either of these diodes is suitable for use as a rectifier capable of rectification of currents in general very nearly equal to the rated current when used as a transistor or at least equal to the maximum base current rating.

In the case of power transistors, and transistors with the collector junction making internal contact with the case, the collector-to-base junction is preferred from practical thermal resistance considerations. The emitter lead can be left open-circuited or it can be tied to the collector.

Germanium transistors like germanium diodes form more efficient rectifiers than silicon transistors and therefore find wider use for rectifying high currents at low voltages.   This is particularly true at high frequencies because of the difficulty of finding suitable rectifiers with low reverse recovery times.

It is worth remembering that in all circuits the same transistors can be used to perform other circuit operations as well as rectification.

Transistors can be used as rectifier diodes in all conventional rectifying circuits including series and parallel operation with the appropriate voltage- or current-sharing components.

Synchronous Rectifiers

Synchronous rectifiers using transistors are desirable for low-voltage and high-current outputs.   The circuits provide good regulation with high efficiency due to low saturation voltage of the transistors.   However, special precautions are necessary to avoid transistor failures.   The transistor must always be either cut off or saturated.   This sets the maximum value of the load current for a given transistor.   Therefore a protection circuit must be incorporated if short-circuit or capacitive load conditions are encountered.

The basic circuit for the synchronous rectifier is shown in Figure 118 (Refs. 111, 112).   The transistor TR_1 operates as a half-wave rectifier on the sinusoidal input.   It is turned off to block one half-cycle of the input, and is saturated during the other half-cycle.

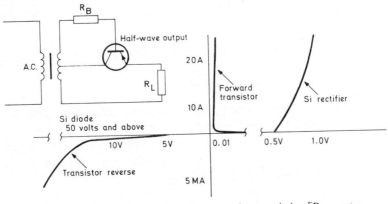

FIGURE 118. (V, I) silicon and synchronous characteristics. [By courtesy of *Electronics Products*, Ref. 111.]

Figures 118, 119, and 120 are 'Reprinted from the April 1964 issue of *Electronic Products Magazine*, Copyright 1964, United Technical Publications, Inc. Div. Cox Broadcasting Corp., Garden City, N.Y., U.S.A.'

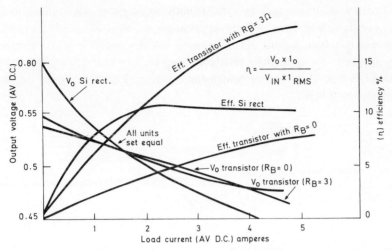

FIGURE 119. Output voltage and efficiency versus load current. [By courtesy of *Electronic Products*, Ref. 111.]

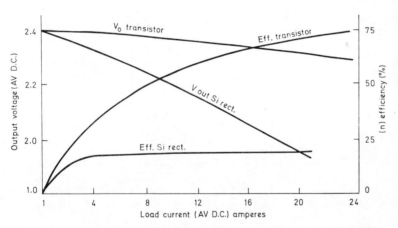

FIGURE 120. Output voltage and efficiency versus load current. [By courtesy of *Electronic Products*, Ref. 111.]

As an example, characteristics of an industrial 25-A silicon rectifier diode and the 2N2728 low-voltage high-current germanium transistor operating as a synchronous rectifier are shown in Figure 118 . The load regulation and efficiencies are plotted in Figures 119 and 120 comparing the synchronous rectifier with a 25-A silicon rectifier diode.

The use of transistors as synchronous rectifiers is not limited to the example described here. Their application to high-frequency circuits is particularly advantageous since, in such circuits, rectifier diodes are not very efficient, owing to their relatively long reverse recovery characteristics as described in this chapter under the heading Rectifier Diodes at High Frequencies (kHz).

CHAPTER 4. INVERTERS AND CONVERTERS

Portable electronic equipment is usually required to operate from a 6, 12, or 24 V battery. For some applications the input voltage is even lower, ranging from 0·8 V to 1·8 V, as in the case of solar cells, thermoelectric cells, etc. In other cases the input voltage may be higher, being 50 V for underground trains or 110 V for railway carriages. However, the supply voltages which are required by the equipment may be of any value from a few volts to several kilovolts.

This is represented diagrammatically in Figure 121 with some readily available battery, shown on the left, and the supply voltage which is needed for an equipment, on the right. It is therefore necessary to use some device which will convert the available voltage to the required value. Such a device, in this case a d.c. converter or inverter, is shown in the middle.

FIGURE 121.   Arrangement for operating equipment from batteries.

There are many methods of converting d.c from low to high voltage using some switching or oscillating device. Transistor d.c. inverters and converters have many advantages over the vibrator and the rotary converter which had irreducible mechanical and electrical losses. D.c. converters can maintain conversion efficiencies of some 70 to 85%, over a wide range of powers, down to a few milliwatts. At high-power levels, from a few watts to several hundred watts, the efficiency may be as high as 90%.

The high efficiency, small size, light weight, and, above all, absence of moving parts make transistor converters very suitable for portable and mobile electronic equipment. They eliminate contact or commutator wear associated with mechanical converters. The need for regular maintenance

185

and the limitation of life therefore disappears, while the absence of sparking and arcing greatly reduces the problems of interference with communications equipment.

There are many possible circuits in which transistors may be used to convert voltages from one value to another. All transistor circuits, however, are either ringing choke or transformer-coupled arrangements.

D.c. converters are circuits which convert a d.c. voltage from one value to a d.c. voltage of a different value. Ringing choke circuits, being followed by a rectified output, are all classed as d.c. converters. Transformer-coupled circuits, however, are basically d.c.-to-a.c. inverters. An a.c. output voltage, whether it is sine wave or square wave, is often used. The transformer-coupled circuits become converters only if they are followed by a stage of rectification before the output is applied to a load.

The production of oscillations need not be performed in the same part of the circuit as that handling the full power. When the oscillator and power stages are separate, the inverter is of the driven type; when they are combined, the inverter is self-oscillating.

The basic operation, together with appropriate design procedures of various circuits, are described in this chapter, and some practical examples are given.

Basic Inverter Principles

All electronic inverters use a system in which the electronic device acts as an interrupter, interruption being the first stage in the conversion process.

In the d.c. converter the transistor is made to oscillate, being alternately bottomed and cut off. This switching action interrupts, or modulates, the current from the d.c. source. The transformer or other voltage step-up device then operates on this fluctuating current, and its output is reconverted to d.c. by rectification.

The oscillator, the only function of which is to convert power from one voltage to another, is primarily required to work reliably and efficiently. The waveform and the operating frequency, although they may be of major importance in some circumstances, can normally be chosen so as to fulfil these primary requirements.

Relaxation oscillators are essentially more efficient than sine-wave oscillators; therefore a form of relaxation oscillator is normally adopted for d.c. converters. This applies particularly when transistors are used.

The dissipation in the junction transistor is very low when it is 'off' and also when it is 'on' and bottomed or saturated. The dissipation is very much higher when the transistor operates unbottomed.   Therefore for the self-oscillating relaxation oscillator to be efficient it requires sufficient positive feedback, from the collector–emitter circuit to the base–emitter circuit, to keep the transistor bottomed for a certain period after switching on.   Regenerative switch-off should take place as soon as the transistor comes out of bottoming, so that the lossy condition, when the transistor is on but not bottomed, is avoided.   Furthermore, as the on characteristic of the bottomed transistor is approximately that of a low resistance, the most efficient form of transistor oscillator, at least so far as dissipation in the transistor is concerned, is one in which the transistor is conducting for as large a part of the operating cycle as possible and the peak-to-mean current ratio is as low as possible.

Several possible oscillator circuits exist but, as they form a vital part of the d.c. converter or inverter, they will be discussed together with the appropriate circuit.

In general, ringing choke circuits are treated as d.c. converters and the transformer-coupled circuits are treated as d.c. inverters.   As mentioned previously, all oscillator circuits are basically d.c. inverters and become converters only if they are followed by a stage of rectification; so a d.c. output is used.

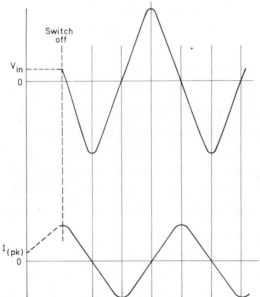

FIGURE 122.   Voltage and current waveforms
after switch-off.

Ringing Choke Converters

Principles of Ringing Choke Circuits

In ringing choke circuits (Ref. 113) no transfer of energy takes place during the on period.   Instead, energy is taken from the battery and stored in the inductance during on period, and is then delivered to the output circuit during the off period as in Figure 122.   The time scale is exaggerated after switch-off in order to show the phase relationship.

The 'On' Period

The equivalent circuit is simply that of Figure 123.   It is assumed that an ideal interrupter, a switch S, is alternately 'made' and 'broken'.   The switch is in series with the d.c. input supply, and, when the switch S is made, a linearly rising current begins to flow, given by $di/dt = V_{in}/L_p$.   After a time t_1 the current flowing is $(V_{in}/L_p)t_1$, and the amount of energy stored is given by

$$E_s = \frac{1}{2} L_p i_{pk}{}^2 = \frac{1}{2} \frac{V_{in}^2}{L_p} t_1^2 \tag{86}$$

If the energy supplied by the battery is considered, another expression for the stored energy is obtained:

$$E_s = V_{in} I_{av} t_1 = \frac{1}{2} V_{in} i_{pk} t_1. \tag{87}$$

where I_{av} is the average current during the time t_1.

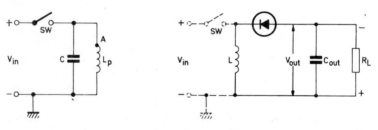

FIGURE 123.   Ringing choke equivalent circuit (left).

FIGURE 124.   Ringing choke with output capacitor and load connected through a diode (right).

The 'Off' Period

If the switch is opened at the time t_1, the current through the inductance' cannot fall instantaneously. If no provision is made for discharging the stored energy, the following happens: the current i_{pk} continues to flow after the break, but it now charges the stray capacitance C of the winding. As the voltage at point A rapidly swings negative, the current decreases, becoming zero when all the stored magnetic energy has left the inductance. The energy is now stored in electrostatic form in the capacitance C, which has been charged through v_{pk} volts, where $\frac{1}{2}Cv_{pk}^2 = \frac{1}{2}Li_{pk}^2$. The capacitance now discharges into the inductance, and an increasing reverse current begins to flow in the inductance. It reaches a maximum when the voltage across the capacitance has fallen to zero. At this point the electrostatic energy has all been converted to electromagnetic energy again, and the current is $-i_{pk}$.

In the ideal case this interchange of energy between capacitance and inductance continues, and sinusoidal oscillations of current and voltage take place as shown in Figure 122, until switch is closed again. In practice the oscillations, which are called 'ringing,' are, of course, damped by the inevitable losses, and they gradually decrease in amplitude and die. The period of the oscillation is given by $2\pi(LC)^{\frac{1}{2}}$.

When a load circuit consisting of a resistor in parallel with a large capacitor is connected through a diode to the ringing choke circuit as in Figure 124, then the ringing is arrested as soon as the diode begins to conduct.

Assuming that the capacitor C_{out} is initially charged to a negative voltage V_{out} (owing, say, to previous cycles of operation), the rapid positive swing of voltage which occurs when the switch is opened is arrested as soon as the voltage across the inductance reaches V_{out}. The transformer voltage is then held at this value while the current flows from the inductance into the capacitance, falling from an initial value approximately equal to i_{pk} to a value $i_{pk} - (V_{out}/L)t$ at rate given by $di/dt = V_{out}/L$, provided that the capacitor C_{out} is so large that the voltage across it is not substantially altered by the flow of charge into it. This linearly decreasing current reaches zero after a time t_2 given by

$$t_2 = \frac{L_p i_{pk}}{V_{out}}. \tag{88}$$

At this point the stored energy has all been transferred to the output circuit. The voltage across the inductance now collapses, and the rectifier is cut off.

If the switch is again closed after being open for a time t_2, the cycle of events will be repeated every $t_1 + t_2$ seconds.

The amount of energy taken from the battery in one second is $\frac{1}{2} V_{in} i_{pk} t_1 \times (t_1 + t_2)$, and, if this is fed into the output circuit without incurring any losses, the output voltage is given by

$$\frac{V_{out}^2}{R_L} = \frac{V_{in} i_{pk} t_1}{2(t_1 + t_2)}, \tag{89}$$

because the output voltage adjusts itself to an equilibrium value at which the energy periodically fed into the output capacitor is steadily dissipated in the load resistance. The output voltage is thus a function of the load.

When the output voltage required is greater than the maximum voltage which can be tolerated across the switching element, the output diode must be attached to an overwinding or a separate winding on the choke, so that the switch voltage is a fraction of V_{out}.

The fact that the output voltage is a function of the load is one of the basic differences between the transformation processes encountered in the ringing choke and the transformer-coupled circuits discussed later.

Basic Ringing Choke Converter Circuit

A basic circuit of the ringing choke converter is given in Figure 125, in which the switch is replaced by a transistor (Refs. 113 to 117).

The principles of operation can be described with reference to the Figure 125 on the assumption that the transistor starts to conduct when the supply is connected.

Because of the collector current flowing in the primary winding there is a voltage induced in the base winding which assists the base current, and owing to the cumulative action the transistor passes from its cut-off state (in which its resistance is high) to its bottomed state (in which its resistance is low). Rapid transistion is made from one state to the other and losses are kept to a minimum.

During the on period the transistor is bottomed, and practically all of the input voltage is applied across the transformer primary. A linearly rising current flows through the transistor and the primary winding. A voltage approximately equal to the supply voltage, V_{cc}, therefore exists

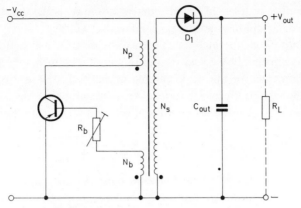

FIGURE 125. Basic d.c. converter circuit.

across the primary winding and a constant voltage of $V_{cc}(N_b/N_p)$ is developed in the base winding producing a constant base current I_B.

The collector current rises until it reaches the value of $h_{FE}I_B$, where h_{FE} is the large-signal current gain of the transistor. At this stage the transistor moves out of bottoming and switch-off begins. The collector voltage rises and the voltage across the primary falls. The voltages across all the windings therefore reverse and the transistor is cut off.

During the off period of the transistor the reverse voltages rise rapidly as ringing commences, but are arrested when the secondary voltage reaches the value of the output voltage. At this point the diode D_1, which so far has been non-conducting, begins to transfer energy from the inductance of the primary to the reservoir capacitor. When the output current has fallen to zero, the rectifier becomes non-conducting, leaving a voltage equal to the output voltage across the choke. The stray capacity of the choke then discharges, causing a reversal of current in the choke and reversal of voltage in the base winding which switches the transistor on again.

The duration of the on period, t_{on}, and the duration of the off period, t_{off}, are given by

$$t_{on} = L_p \frac{I_{CM}}{V_{cc}}. \tag{90}$$

$$t_{off} = L_p \frac{I_{CM}}{I_{out}} \frac{N_s}{N_p} \tag{91}$$

where L_p is the inductance of the primary winding, I_{CM} is the peak collector current of the transistor, which is the magnetising current of the transformer,

N_p is the number of turns of the primary winding, N_s is the number of turns of the secondary winding, and V_{out} is the output voltage.

The circuit thus operates as a relaxation oscillator generating a rectangular waveform. It is based on the ringing choke principle, but ringing is arrested long before the peak voltage is reached. High efficiency and power-handling capacity are obtained by arranging that the transistor is bottomed when it is in the on condition, that is by ensuring that

$$I_C < h_{FE}I_B, \qquad (92)$$

where I_C is the collector current, I_B the base current, and h_{FE} the static current gain with grounded emitter. The operating characteristics of the transistor are given in Figure 126, and the waveforms for one complete cycle in Figure 127.

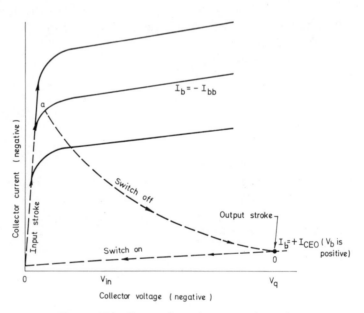

FIGURE 126.   Locus of transistor operating point.

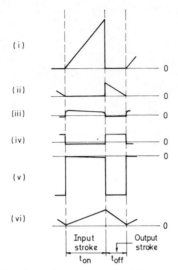

FGURE 127.   Waveforms in basic circuit: (i) input current; (ii) current in secondary; (iii) base current (negative); (iv) base voltage; (v) collector voltage; (vi) flux in transformer core.

Design Considerations

The design of the transformer is normally based on the available supply voltage, the required output voltage and current, and the required physical size of the converter.

Other factors affecting the design are (i) operating frequency, (ii) core material, (iii) air gap, (iv) number of primary turns, and (v) turns ratios of the secondary and the base windings to the primary.

Operating Frequency

A correct choice of operating frequency is desirable for high efficiency and low transistor dissipation.   The choice is also affected by the core material used and by the fact that smoothing and input decoupling are easier at high frequencies.

The variation of the principal energy losses with frequency are shown in Figure 128.   The transient loss in the transistor increases with frequency. It is therefore necessary either to use high-frequency transistors or to restrict the operation to lower frequencies, as in the case of germańium power transistors.

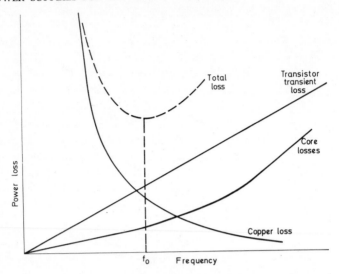

FIGURE 128. Variation of power losses with frequency.

The curve of total losses shows that there is an optimum frequency f_o for a maximum overall efficiency. For a maximum power output for a given transistor, usually a much lower frequency is necessary. Therefore a compromise value below f_o is chosen, which for germanium power transistors is found to lie between 100 Hz and 10 kHz. In the case of silicon power transistors, owing to much higher f_T, the optimum frequency f_o is between 5 kHz and 50 kHz.

Core Material

For low and medium-power ringing choke converters the frequency of the operation can be above 1 kHz and a Ferroxcube core is used. For high-power converters a lower frequency is necessary.

A high saturation flux density is required to allow operation at low frequencies with low copper and transient losses. Laminated iron cores can be used in most cases. However, grain oriented 'C' cores are preferred owing to higher efficiency and a smaller physical size.

Peak Current, Output Voltage, and Power

If losses are neglected, the output voltage at an operating frequency is given by

$$\frac{V_{\text{out}}^2}{R_L} = fL_p I_{CM}^2, \tag{93}$$

where $\frac{1}{2}L_p I_{CM}^2$ is the energy delivered to the reservoir capacitor in each cycle.

The output power

$$P_{\text{out}} = \frac{V_{\text{out}}^2}{R_L} \tag{94}$$

or

$$P_{\text{out}} = \eta V_{cc} I_{IN}. \tag{95}$$

where η is the conversion efficiency and I_{IN} is the average supply current. But

$$I_{IN} = \frac{I_{CM}}{2} \times \frac{t_{\text{on}}}{T} ,$$

where $T = t_{\text{on}} + t_{\text{off}}$ is the period of the oscillation.   Therefore

$$P_{\text{out}} = \eta V_{cc} \frac{I_{CM}}{2} \times \frac{t_{\text{on}}}{T}. \tag{96}$$

From Equation (96)

$$I_{CM} = 2 \frac{T}{t_{\text{on}}} \times \frac{P_{\text{out}}}{V_{cc}\eta}. \tag{97}$$

For an assessment of the peak current, which is needed for choosing a suitable transistor, the following two assumptions are made.   On the assumption that

$$\frac{t_{\text{on}}}{T} = \tfrac{2}{3} \text{ and } \eta = \tfrac{3}{4},$$

then

$$I_{CM} = 4 \frac{P_{\text{out}}}{V_{cc}}$$

$$= 4 \frac{I_{\text{out}} V_{\text{out}}}{V_{cc}}. \tag{98}$$

Primary Winding

The primary current of the transformer is the magnetising current, if the small fraction due to the feedback current is neglected. The primary inductance therefore is given by

$$L_{p} = V_{p} \frac{t_{on}}{I_{CM}}. \tag{99}$$

where V_{p}, the voltage across the primary, is

$$V_{p} = V_{cc} - V_{CE(sat)} - V_{Rp}, \tag{100}$$

$V_{CE(sat)}$ being the saturation voltage of the transistor and V_{Rp} being the voltage drop across the resistance of the primary winding.

Equation (99) can also be written in a form

$$L_{p} = \frac{V_{p}}{I_{CM}} + \frac{1}{f} + \frac{t_{on}}{T}. \tag{101}$$

The number of the primary turns is given by

$$N_{p} = \frac{I_{CM}L_{p} + 10^{8}}{\cdot A + B_{M}}. \tag{102}$$

where A is the cross-sectional area of the core in square centimetres and B_{M} is the peak value of flux density in gauss. Equation (102) is a general equation which is used in conjunction with iron cores. In the case of ferrite cores more accurate data are available from the manufacturers and a much simpler equation is used to arrive at the number of the primary turns, which in this case is given by

$$N_{p} = \alpha (L_{p})^{\frac{1}{2}} , \tag{103}$$

where α is the number of turns for 1 mH and L_{p} is the inductance of the primary in millihenries.

Feedback Winding

Having decided on the primary turns the number of turns required for the base winding to provide sufficient feedback for oscillation can be found from

$$N_f = N_p \frac{V_f}{V_p}, \qquad (104)$$

where V_f is the feedback voltage required for the oscillation which is

$$V_f = B_{BE} + V_{RB}, \qquad (105)$$

where V_{BE} is the base–emitter voltage required for the peak collector current given by Equation (97) and V_{RB} is the voltage drop across the external base resistor, together with the voltage drop across the resistance of the feedback winding.

Secondary Winding

The output voltage is determined by the factors given in Equation (93) and not by the primary-to-secondary turns ratio.

Neglecting the ripple voltage across the reservoir capacitor, the peak secondary voltage is given by

$$V_s = V_{out} + V_d + V_{RS}, \qquad (106)$$

where V_d is the forward voltage drop across the rectifier diode and V_{RS} is the voltage drop across the resistance of the secondary winding.

The peak voltage which then appears across the collector–emitter junction. $V_{CE(pk)}$, when the transistor is cut off is

$$V_{CE(pk)} = V_{cc} + (V_{out} + V_d + V_{RS}) \frac{N_p}{N_s}. \qquad (107)$$

Since it is necessary to keep $V_{CE(pk)}$ within the maximum rated peak collector voltage V_{CEM}, in order to avoid avalanche breakdown the maximum number N_s, of secondary turns from Equation (107) is given by

$$N_s = N_p \frac{V_{out} + V_d + V_{RS}}{V_{CEM} - V_{cc}} \qquad (108)$$

It is possible to use a higher number of secondary turns, but then a higher peak collector current is required owing to the reduced t_{on}/T ratio. As a result of which, the efficiency may be affected owing to the increased copper and transient losses. Also the rectifier inverse voltage increases with the increased secondary turns. Therefore the number of turns of the secondary winding should not exceed the turns obtained from Equation (108) by more than is necessary for a safety margin.

Output Smoothing Capacitor and Decoupling

The peak ripple voltage superimposed on V_{out} is given by

$$\Delta V_{out} = \frac{V_{out} t_{on}}{R_L C_{out}}. \tag{109}$$

If C_{out} is moderately large and the operating frequency is fairly high (thus giving a small value of t_{on}), then the ripple voltage can be quite small. In general the output capacitor should have a value such that

$$R_L C_{out} \geqslant 10 t_{on}. \tag{110}$$

Substituting for R_L from Equation (94) and rearranging, the following is obtained;

$$C_{out} \geqslant \frac{10 t_{on} P_{out}}{V_{out}^2}. \tag{111}$$

The a.c. component of the input current develops a ripple voltage across the internal resistance of the battery. If other equipment is operated from the same battery, it may suffer interference from this ripple voltage. For operation at high frequencies, further precautions are necessary as the ripple voltage will cause radiation from the connecting leads. Decoupling of the converter input will then be necessary. The resistance of any chokes used in the decoupling filter should be kept low, so that the overall efficiency is not unduly affected.

Initiation of Oscillation

If the input voltage is applied gradually when the converter is first switched on, the cycle of operation will not necessarily start. Some forward bias should therefore be applied to the base of the transistor to reduce the input resistance and to increase the gain of the feedback loop to the point of instability and regenerative switch-on. When an appreciable output voltage has been built up, the oscillations are maintained by the base voltage swing. In this condition the bias can be removed.

The forward-bias voltage (which should be about 200 mV for germanium transistors and 600 mV for silicon transistors) can be provided by means of a voltage divider connected across the converter input (Figure 129). The bias voltage is in series with the base winding of the transformer. The voltage should not be higher than is necessary to provide reliable starting, and it should preferably be applied only until oscillation is fully established. Permanent bias can prevent the operation of the overload protection circuit described on page 200.

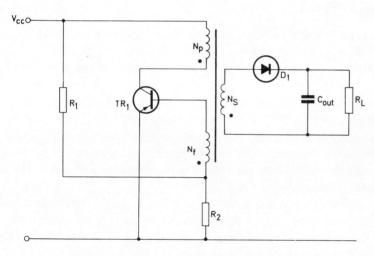

FIGURE 129. Ringing choke inverter with starting bias.

The initial switch-on can be complicated by 'bouncing' of the input switch. The input voltage during bouncing can be maintained by the connection of a capacitor (say 50 μF) immediately across the converter input on the converter side of the switch.

Heavy loading of the output can also make starting difficult. The static condition $V_b = 0$, $V_c = V_{cc}$ is moderately stable when the transistor is substantially off, because the loop gain of the oscillator is then low and the base input resistance is high. Heavy transformer loading accentuates this stability; therefore for reliable starting the output load should initially be as low as possible. The loading is increased by stray capacitances across the transformer and by the capacitance of the semiconductor rectifier (which is greatest in the absence of inverse voltage).

The above considerations hold when the input voltage is applied gradually. Sudden application, on the other hand, produces shock excitation of the self-resonant circuits, and the base can swing sufficiently negative for the assistance of bias to be unnecessary.

But here again the switch must make cleanly (that is there should be no bouncing), and the output loading should be not too heavy at the moment of switch-on. Converters heavily loaded and converters with voltage multipliers usually require a permanent starting arrangement.

Protection Circuit

A protection circuit is required if a ringing choke converter is to operate when the load is removed even for a very short period. This is because, under open-circuited load conditions, high voltages are generated which appear across collector–emitter and base–emitter junctions in the reverse direction. These voltages may be two to three times the values permitted for normal operation. Sometimes delayed load application may provide similar conditions. High-tension supplies for thermionic valves are an example The converter operates unloaded while the heaters warm up, being driven directly from the battery to avoid loss in efficiency and the use of more powerful transistors.

A simple protection circuit is shown in Figure 130. During normal operation the diode D_2 is reverse-biased by the supply voltage V_{cc} and does not conduct. The winding N_f is arranged so that the voltage induced is not sufficient to overcome the bias provided by the battery. The diode D_2 conducts, however, when the ouput voltage rises above the normal value which makes the induced voltage in the winding N_f higher than V_{cc}. D.c. voltage is then developed across $C_2 R_2$ which tends to bias off the transistor.

The reverse collector voltage is then limited to a value $V_{cc} + V_{cc}(N_p/N_f)$. Good coupling is necessary between N_p and N_f which may be best realised by bifilar coupling.

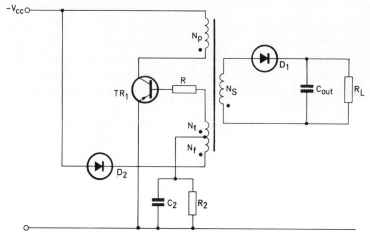

FIGURE 130.   Ringing choke inverter with over voltage and open-circuit load protection circuit.

Another example of operation under open-circuited load conditions immediately after switch-on is in the case of a fluorescent lamp inverter. Ringing choke inverters are often used for such applications because of the economic solution which they provide. The circuits designed very often meet the reverse base–emitter voltage rating for normal load operation, but not for open-circuited load conditions, particularly in the case of silicon transistors. This results in operation outside the normal published data for the transistors and may lead to catastrophic failure during initial testing or, if a powerful enough transistor is used, gradual deterioration will result in greatly reduced transistor life. The so-called *soak tests* are not a proof of the equipment reliability, but deliberate avoidance of the published data. Designers should ensure that all tests including normal load and unloaded conditions should not exceed the published data at any time. It is the designer's duty to provide sufficient safety margin so that under adverse conditions of the supply voltage variation, including temperature changes and loading limits, there is enough safety margin for switch-on transients, switch-off transients, and any other conditions which may be encountered during normal operation.

Other Ringing Choke Circuits

Common Collector Ringing Choke Circuit

It is often advantageous to use the transistor in the common collector configuration to reduce mounting problems connected with insulation of the transistor from the heatsink. The transistor can then be bolted directly onto the heatsink, forming a part of the common chassis of the equipment, thus improving thermal conductivity between the heatsink and the transistor, and at the same time reducing the overall cost of the equipment.

A basic common collector circuit is shown in Figure 131.

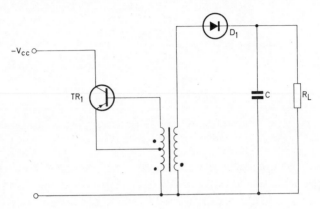

FIGURE 131.   Basic common collector ringing choke
converter.

If the isolation between input and output circuits is not necessary an auto-transformer can be used. The auto-transformer provides better utilisation of the available winding space and therefore results in a smaller size of the transformer.

Detailed analysis of the common collector circuit, which equally well applies to other types of the ringing choke converter, can be found in Refs. 118, 119.

The common collector ringing choke converters, as shown in their basic form in Figures 131 and 132, may not start to oscillate when the supply voltage V_{cc} is applied. Starting bias can be provided by connecting the resistors R_1 and R_2 as shown in Figure 133. The resistor R_3 is added to stabilise the oscillation and to reduce the effect of the spreads in h_{FE} of the transistor.

Furthermore, the converter circuit can be rearranged to enable the collector and the heatsink to be connected to the common earth as shown in Figure 134.

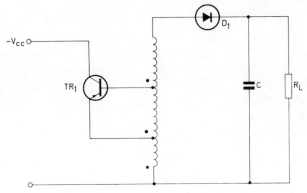

FIGURE 132.   Common collector ringing choke circuit using auto-transformer.

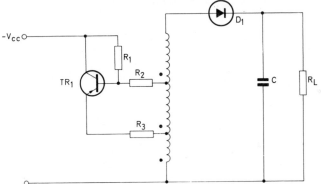

FIGURE 133.   Starting bias arrangement for Figure 132.

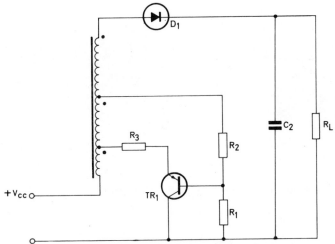

FIGURE 134.   Arrangement for the connection of the collector
and the heatsink to common earth.

Converters With the Load in the Base Circuit of the Transistor

D.c. converter circuits in which the load current is at the same time the base current, or part of the base current, are shown in Figures 135 and 136. These circuits operate partly as transformer-coupled and partly as ringing choke circuits with heavy capacitive loading which damps the overswing and limits the reverse base–emitter voltage to the value acceptable by the transistor. The voltage developed across the load resistor R_L is due to the transformer coupling, but the transistor re-set is a result of the ring derived from the energy stored in the transformer.

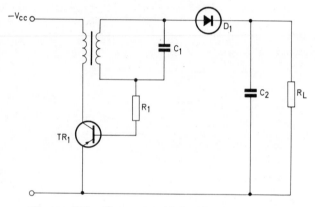

FIGURE 135. Converter with load in the base circuit.

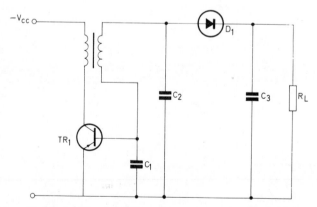

FIGURE 136. Another arrangement for connecting R_L in the base circuit.

The circuits are used in conjunction with germanium transistors whose leakage current is very often sufficient to initiate oscillation without having to use special starting arrangements.

In the case of silicon planar transistors, the circuits lose their usefulness because not only the starting circuits are necessary, but also their use is limited to relatively low output voltage levels because of the low reverse base–emitter voltage rating.

Transformer-coupled Inverters and Converters

To stress the importance of the difference between the ringing choke converter and the transformer-coupled circuits the principles of the latter are now outlined.   This is followed by a description of those circuits which, in addition to being transformer-coupled, are also push–pull arrangements.

The methods of push–pull operation include (i) sinusoidal class B and class C circuits, (ii) square-wave driven output stages, and (iii) square-wave self-oscillating systems (Refs. 113, 120 to 137).

There are three basic functions to be performed in a square-wave self-oscillating inverter: (a) oscillation, (b) power transfer, and (c) frequency control.

The oscillation is obtained by feeding a fraction of the output voltage back to the input and by arranging the loop gain of the stage to be greater than one.   The power transfer is then achieved by means of a transformer coupling to the oscillating circuit.

The frequency control and the mode of operation of these circuits depend on whether the transformer is designed to saturate or not, and on whether the collector circuit or the base drive is the controlling factor. This in turn means that the frequency, or switching rate, is controlled either at high-power level in the collector circuit, or at low-power level in the base circuit.   The classification of transformer-coupled inverters is shown diagrammatically in Figure 137.

Inverters with single saturable transformers belong to a family of circuits controlled at high-power level.   Other arrangements, namely (a) inverters with two transformers, (b) inverters with CR timing, and (c) inverters with LR timing and inverters with LC tuned circuit timing, belong to a family of circuits controlled at low-power levels utilising the base current as the controlling element.

In addition to requiring less power for controlling purposes and therefore being more efficient, the base controlled circuits have a further advantage

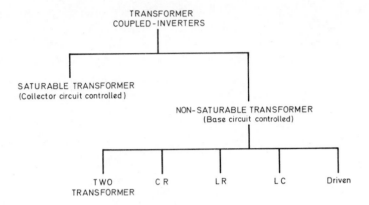

FIGURE 137.   Classification of transformer-coupled inverters.

in that they can easily be converted to driven type inverters merely by dis-
connecting the feedback circuit and by applying external drive to the base
circuits.

The transformer-coupled inverters and converters also embrace bridge
type configurations which are particularly useful for high input voltages.
The classification of transformer-coupled inverters of Figure 137 applies
therefore to both push–pull and bridge circuits.

Principles of Transformer- coupled Circuits

In transformer-coupled circuits (Ref. 113) the transformation of energy
takes place while the current is flowing in the primary winding of the transformer after an application of the supply voltage.

A circuit with an ideal transformer is shown in Figure 138.   It is assumed
that an ideal interrupter (a switch is alternately 'made' and 'broken') is
in series with the d.c. input supply.   A load R_L, in series with a diode D,
appears across the output of the transformer.   When the switch S is made,
the whole of the input voltage V_{in} appears across the primary and a voltage
$V_{in}(N_s/N_p)$ appears across the secondary.

The secondary voltage causes a current $(N_s/N_p)(V_{in}/R_L)$ to flow into the
load through the diode D (which is connected so as to conduct during this
part of the cycle).   The action on the primary side during this period is
best seen if the equivalent circuit shown in Figure 139 is used, leakage
inductance and winding resistances being neglected.

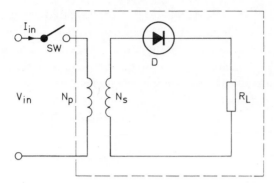

FIGURE 138. Transformer-coupled circuit.

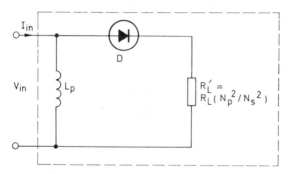

FIGURE 139. Equivalent circuit of Figure 138.

When the switch S is made, a constant current I_r, equal to V_{in}/R_L', will flow into the 'load equivalent' R_L', and a linearly rising current I_1 (starting from zero if there has been no previous current flow in the inductance) will flow in the shunt inductance L_p. This is the magnetising current, its rate of rise being given by $di/dt = V_{in}/L_p$. Thus the input current waveform, up to a time t_1 when the switch S is opened, is that shown in Figure 140(a). The shaded part is the component of input current which has been transferred to the load, whereas the dotted part is the current I_1 which builds up magnetic flow in the transformer core as shown in Figure 141(a). If all ordinates in the plot of currents (Figure 140(a)) are multiplied by V_{in}, a diagram of power flow against time is obtained (Figure 142(a)). The shaded area represents the energy which has been transferred to the load, and the dotted area the energy which has gone into establishing the final flux level in the inductance of the transformer and is thus stored in magnetic form. If an initial current equal to $-i_{1(pk)}$ and an initial flux equal to $-\phi_{pk}$ exist in the transformer primary (as is the case in some symmetrical push–pull circuits), then Figures 140(a), 141(a), and 142(a) are replaced by Figures 140(b), 141(b), and 142(b) respectively. In Figure 142(b) the shaded area still represents the energy transferred into the load, and the dotted area the energy stored at the end of the conduction period, but the area which is both shaded and dotted represents energy which was stored in magnetic form in the transformer before the switch was closed and which was then recovered and transferred to the load during the first half of the conduction period.

Energy Transfer

The amount of energy transferred to the load in each conduction period of duration t_1 is given by

$$E_t = V_{in} I_r t_1 = \frac{V_{in}^2}{R_L'} t_1$$

$$= \frac{V_{in}^2}{R_L} \left(\frac{N_s}{N_p}\right)^2 t_1 = \frac{V_{out}^2}{R_L} t_1. \tag{112}$$

and the energy remaining stored at the end of the conduction period (where the initial current in the primary is zero) is given by

$$E_s = \tfrac{1}{2} L_p i_{1(pk)}^2 = \frac{1}{2} \frac{V_{in}^2}{L_p} t_1^2. \tag{113}$$

When the initial current in the primary is $-i_{1(pk)}$, then

$$E_s = \tfrac{1}{2} L_p i_{1(pk)}^2 = \frac{1}{2} \frac{V_{in}^2}{L_p} (\tfrac{1}{2} t_1)^2$$

$$= \frac{1}{8} \frac{V_{in}^2}{L_p} t_1^2. \tag{114}$$

In this case the net storage of energy during the whole period is zero. The ratio of energy stored to energy transferred is, in both cases, proportional to the following expression

$$\frac{R_L'}{L_p} t_1.$$

Conditions During the 'Off' Period

When the switch is opened, the power source is disconnected and the flux in the inductance must decrease; therefore the voltages across all the windings reverse, the rectifier D is biased in its blocking direction, and R_L is thus disconnected. The flux level in the inductance must be restored during the off period to its initial value in readiness for the next cycle; therefore the stored energy must be dissipated or made use of in some way during the off period.

The simple circuit of Figure 138 is unsatisfactory because no means of dealing with the stored energy exists. When the switch is opened and primary current ceases to flow, the magnetic field in the inductance collapses extremely rapidly, and a very high voltage appears transiently across the switch and across the diode in its blocking direction. This high voltage could damage the switch (which in practice is a transistor) and the diode. Some method of preventing this must therefore be added to the simple circuit.

Methods of Dealing with the Stored Energy

There are several alternative methods of dealing with the stored energy; (i) by connecting a resistor across one of the windings, (ii) by connecting a capacitor across one of the windings, (iii) by feeding the stored energy to the load, (iv) by returning the stored energy to the input circuit.

These methods will be examined critically to see how effective they are (a) in preserving good overall efficiency, and (b) in limiting the voltages occurring across the transformer during the 'off' period. The latter consideration is of great importance in practice, as the voltages across the switching transistor and the diode must be kept within the ratings of these components.

Method (i). A resistor R_d across, say, the primary of the transformer will dissipate the stored energy. An exponentially decreasing current, initially equal to $i_{1(pk)}$, will flow in R_d at the beginning of the off period, and the voltage $i_{1(pk)}R_d$ generated across the primary will depend on the value of R_d. In order to keep the voltage to a permissible value, R_d must be chosen to be less than a certain value which will normally be comparable with the load equivalent resistance $R_L{}'$. Thus the use of such a dissipative resistor is inefficient not only because the stored energy is lost, but also because a substantial amount of energy is taken from the battery during the 'on' period. The drain on the battery can be avoided by inserting a rectifier in series with R_d, but even then the circuit is inefficient, and the rectifier could be employed to better advantage—for example, as in methods (iii) and (iv). Another disadvantage of this method is that the discharge of stored energy is relatively slow.

Method (ii). A capacitor C can be connected across a transformer winding. If it is connected across, say, the primary, it will be charged to V_{in} during the on period. When the switch is opened, ringing will take place. The current $i_{1(pk)}$ previously flowing in the shunt inductance from the battery now flows into C, decreasing sinusoidally until the energy which was stored in magnetic form is all converted to electrostatic energy of charge in the capacitor   The capacitor will then discharge into the inductance until all the energy is once more in magnetic form, and so on. The relevant waveforms are given in Figure 122. The peak voltage across the capacitor is given by

$$\tfrac{1}{2}CV^2 = \tfrac{1}{2}L_p i_{1(pk)}{}^2.$$

When the permissible peak voltage (which appears in series with V_{in} across the switch) is low, the value of C must be correspondingly high.

This method of dealing with the excess energy has the advantage that, after exactly one half-cycle of ringing, a current of $-i_{1(pk)}$ is flowing in the inductance. If the switch is closed again at this point, the conditions shown in Figures 140(b), 141(b), and 142(b) prevail, and the stored energy is transferred to the load during the first half of the on period. As the flux swing is bi-directional and nearly symmetrical, there is the additional advantage that the whole of the (B, H) loop of the core is used and the pre-magnetisation is small.

On the other hand, the capacitor circuit has the following disadvantages: (a) if switching does not take place exactly at the right instant, high charging current flows through the switching transistor and it is difficult to keep the transistor bottomed; (b) the reversal of the transformer voltages is greatly slowed down by the capacitor, and the transistor cannot be turned on and off quickly in a self-oscillating circuit if it is controlled by a voltage from the transformer. This latter effect (as will be shown later) causes a substantial power loss in the transistor.

Method (iii). The stored energy can be fed to the output load if a rectifier is connected from a transformer winding to the load so that the rectifier conducts during the off period. A circuit with a separate winding for this purpose is shown in Figure 143.

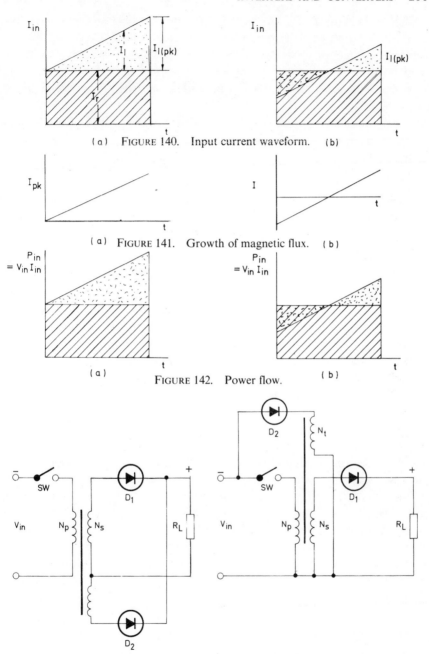

(a)   FIGURE 140.   Input current waveform.   (b)

(a)   FIGURE 141.   Growth of magnetic flux.   (b)

(a)

FIGURE 142.   Power flow.   (b)

FIGURE 143.   Circuit for returning energy stored to the load (left).
FIGURE 144.   Circuit for returning energy stored to the input (right).

The operation of the circuit is similar in principle to method (i), but with the advantage that (neglecting any losses which occur during the recovery of the stored energy) the whole of the energy taken from the battery is supplied to the load.   The voltage across the load is far from constant over the cycle, but a large capacitor in parallel with it will provide adequate smoothing.

This circuit has one serious disadvantage: it loses the main advantage of the transformer step-up principle, namely the obtaining of a constant output voltage irrespective of the value of the load.   Whenever stored energy is fed to the output circuit, the output voltage will depend on the load value. This is shown by the following argument: the amount of energy stored per cycle is independent of the value of the load (from Equations (113) and (114). If a constant amount of energy is fed to a varying load, then the average voltage developed across the load must vary with the load.

The circuit thus has a larger output impedance than the other transformer-coupled circuits, though the impedance is lower than that of the basic ringing choke circuit.   As the latter is simpler it will normally be preferred when a high output impedance is admissible.

Method (iv).   The stored energy can be returned to the input circuit from a transformer winding in series with a rectifier.   This arrangement (which is the most generally useful for transformer-coupled converters using a single transistor) is shown in Figure 144.   It will be shown later that push–pull circuits, though apparently dissimilar, in fact use the same mechanism for returning the stored energy to the input.

In Figure 144 the rectifier D_2 is non-conducting during the on period. When the switch S is opened, the transformer voltages reverse and rise rapidly.   As soon as the voltage across N_t reaches the input voltage, however, the rise is arrested and current flows from the transformer through the rectifier D_2 to recharge the input battery.   The initial value of the current is $i_{1(pk)}(N_p/N_t)$, and it decreases linearly at the rate $di/dt = V_{in}/L_t$, where L_t is the inductance of the winding N_t.   When the current has fallen to zero, the switch can be made again and the cycle repeated   A large capacitance should be connected across the input voltage source where it is undesirable to inject current into it, or when it has a high internal impedance.

General Observations

In all the above circuits, with the exception of Figure 143, the output voltage is governed only by the input voltage and the primary-to-secondary step-up ratio. The regulation of transformer-coupled converters is thus intrinsically good.

In the single-transistor circuits an output is obtained only during the on periods, but it can be maintained over the whole cycle by connecting a large capacitor across the load. When such a capacitor (which is peak-charged during the on period) is present, the operation of the circuit and the waveform of the input current are modified, but a good regulation is maintained.

It is always desirable to keep the amount of stored energy relatively low, because it is then easier to deal with it. (For example, the peak voltages developed in arrangements (i) and (ii), and the current ratings of the rectifiers D_1 and D_2, depend on it.) The stored energy can be minimised by keeping the ratio R_L'/L_p as small as is consistent with low primary winding resistance and low switching losses.

Single-ended Transformer-coupled Circuits

Conventional Single-Ended Circuits

The two circuits shown in Figures 145 and 146 are single-ended versions of the transformer-coupled circuits (Ref. 113). The diodes D_2 are required to return the stored energy to the input circuit as discussed on page 179.

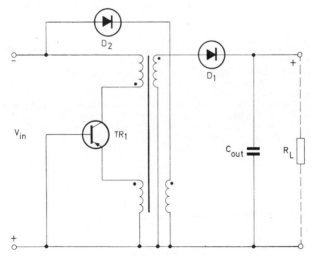

FIGURE 145. Common base single-ended transformer-coupled circuit.

Energy restoration takes place whatever the number of turns on the winding in series with D_2. When the transistor is cut off, the voltage across the winding rises to the battery voltage plus the forward drop of the diode plus the voltage drops in the winding and across the internal resistance of the battery. A corresponding voltage is induced in the primary circuit. The number of turns in the energy return winding must be chosen so that the peak collector voltage rating of the transistor is not exceeded.

The circuit shown in Figure 146 is similar to the ringing choke converter with reduced output impedance shown in Figure 125. It differs structurally only in the turns ratio N_s/N_p required for a given set of conditions, and in the sense in which the secondary is connected. In practice the transformer-coupled circuit will usually have a better regulation than the ringing choke circuit (Figure 125), but its efficiency will be lower. Other forms of single-ended circuit will be suggested by the discussion of transformer-coupled circuits on pages 209-212. The circuit shown in Figure 146 can be rearranged for the common collector connection.

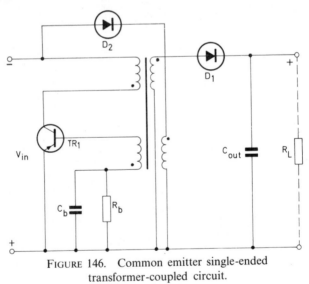

FIGURE 146.   Common emitter single-ended
transformer-coupled circuit.

New Single-ended Circuits

The disadvantages of the conventional ringing choke inverter circuit shown in Figure 124 are that high-voltage and high-current spikes are generated every time the transistor is switched off. This is even more apparent when the circuit is used with an inductive load such as is encountered in the case of fluorescent lamp circuits and in particular with circuits using the leakage type of transformer which acts as a ballast for the fluorescent lamp and at the same time provides sufficient drive for reliable starting of the lamp. The most dangerous manner of operation of such a circuit is under no load condition giving high base–emitter voltage swing in the reverse direction. This is disastrous for silicon planar transistors and therefore the ringing choke circuit in its normal form is not suitable for them.

The conventional single-ended circuits require energy return circuits with close coupling between the various windings of the transformer and therefore are not suitable with leakage types of transformers.

The new circuit (Ref. 138) overcomes these disadvantages and provides safe operation free from high-voltage spikes irrespective of the type of the transformer used.

Linear, saturable, or leakage types of transformers can be used as required for a particular application.

The new circuit in its basic form is shown in Figure 147. The transistor TR_1 is connected to the primary of the transformer N_p. The secondary winding N_s of the transformer performs two functions: it provides the means for controlling the oscillation, and at the same time provides the necessary coupling to the primary in order to obtain the required output voltage.

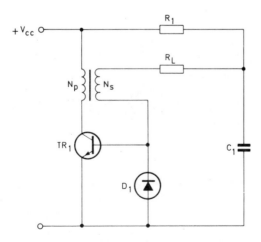

FIGURE 147. Basic single-ended inverter.

It will be observed that the load current is at the same time the base current. Therefore, when the load is removed, the base is open-circuited and no oscillation occurs. Thus automatic open-circuit load protection is built in, with an additional advantage that there is no danger of high voltage being generated under no load condition, which is the case in the ringing choke inverters and converters.

The diode D_1 in Figure 147 provides transistor protection against reverse base–emitter voltage during normal operation, and the capacitor C_1 provides a means of adjusting the frequency of the oscillation and at the same time provides a return path for the load current.

The resistor R_1 is a starting resistor which allows initial base current for the circuit to commence its operation.

It is possible to operate the single-ended inverter in any other of the available transistor connections (Ref. 138). The common collector configuration of the single-ended inverter is shown in Figure 148.

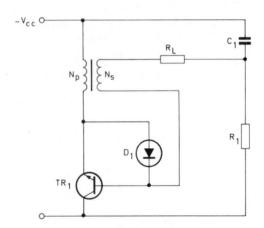

FIGURE 148. Common collector configuration of single-ended inverter.

Figure 149 shows the single-ended inverter with transformer connections capable of supplying multiple outputs. This can be further extended so that rectified output voltages or a mixture of a.c./d.c. outputs may be obtained as in Figure 150. Here the load R_{L1} is supplied with a d.c. voltage developed across the capacitor C_1, and R_{L2} is supplied with a.c. The capacitor C_2 is used to adjust the frequency of operation.

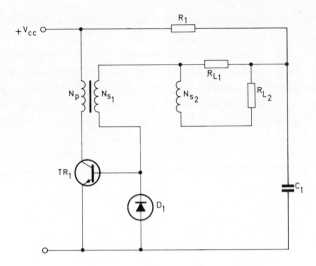

FIGURE 149. Single-ended inverter for multiple outputs.

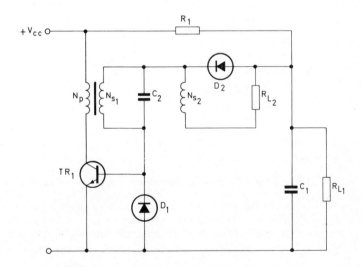

FIGURE 150. Single-ended converter/inverter circuit.

It should be noted that the circuit shown in Figure 150 will oscillate if either R_{L1} or R_{L2} or both loads are open-circuited. The capacitor C_2 in this case serves the purpose of limiting the peak voltage swing across the secondary winding N_{s1}.

Similarly the circuits shown in Figures 147 and 149 can be made to oscillate under no load conditions by connecting the resistor R_1 and the capacitor C_1 onto the secondary winding side of the load resistor R_L. This type of operation, although possible, is not desirable unless it is required to connect one side of the load in common with the supply line, or to a common earth. The value of R_1 should then be adjusted so that the circuit has just sufficient gain for reliable operation under full-load condition so that no excessive dissipation occurs under the no-load condition.

The most useful application of the new single-ended circuit (Ref. 138) is for fluorescent lamp operation with a leakage type of the transformer (Figure 151)

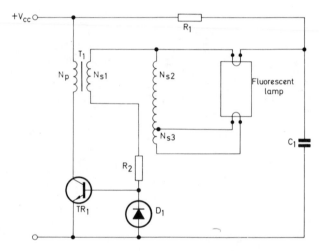

FIGURE 151.   Single-ended fluorescent lamp inverter.

It is known that the operation of inverter circuits with leakage type of output transformers has disastrous effects on the transistors, in particular in the case of a ringing choke type of inverter. The high leakage inductance required for lamp ballasting and the high energy stored in the transformer generate high-voltage and high-current spikes after switch-on and before the lamp strikes. In addition operation under no-load conditions is particularly dangerous, when planar transistors are used. The reverse voltage applied to the base–emitter junction is inevitably exceeded causing *channelling* to occur for the duration when clamping is effective prior to the breakdown of the transistor.

Protecting the junction with a diode is not usually possible as it would limit the required voltage swing across the secondary which would make the starting of the lamp rather difficult.

In the new circuit, however, the operation is modified since the base current is limited, not by the available drive, but by the base current required for the load. The energy stored in the transformer during the conduction period of the transistor is transferred to the load at a rate determined by the resistive value of the load during the cut-off period of the transistor, and the base–emitter junction is then protected by means of the diode D_1. Open-circuiting the load, by withdrawing the lamp, stops the base current immediately so that the circuit cannot possibly oscillate.

If the load is interrupted at the peak value of the collector current, that is in the maximum stored energy condition, then the diode D_1 will clamp the base–emitter junction of the transistor and share the energy dissipation preventing any voltage or current spikes to occur. The only important diode rating is that it should withstand the peak energy during one half-cycle of operation. Under normal load conditions the diode is subjected to a much lower strain as the bulk of the energy is transferred to the load. If operation at high frequency is required, then the switching characteristics of the diode should be considered, in particular the reverse recovery time.

Finally, if it is required for the transistor to be connected to a common earthing line, thus easing the cooling problem by direct connection to the heat sink, a common collector version is shown in Figure 152.

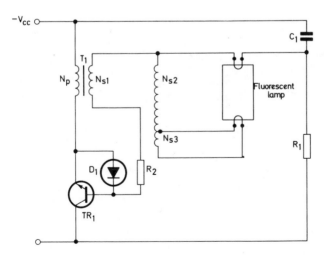

FIGURE 152.   Common collector version of Figure 151.

An additional advantage of the new circuit is that it operates over a very wide range of supply voltages. A change of $\pm$ 50% can be tolerated provided it is within the transistor ratings.

Push–Pull Inverters

All push–pull inverters are basically transformer-coupled circuits (Refs. 113, 120 to 141). For high efficiency, the systems used are of the square-wave oscillating type—there are three well-known configurations of these: (a) common base, Figure 153, (b) common collector, Figure 154, and (c) common emitter, Figure 155. Each circuit can be made into a self-oscillating relaxation oscillator in which transistors are operated either in the cut-off or in the bottomed condition. Since the common emitter arrangement is the most efficient, and therefore the most commonly used, the discussion will be limited to the various forms of this type of circuit. The basic principles, nevertheless, apply to all three connections.

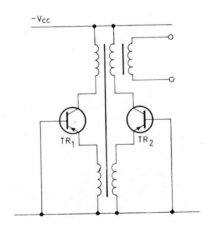

FIGURE 153. Common base.

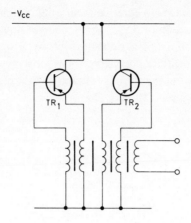

FIGURE 154.   Common collector.

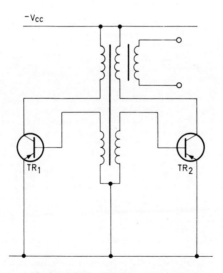

FIGURE 155.   Common emitter.

A reference is made to the conventional common emitter inverter (Ref. 126) shown in Figure 155, which uses a single transformer. The transformer can be either of a non-saturable type or of a saturable type. The circuit operation differs according to which type of the transformer is used. The difference in operation is explained with reference to the collector currents shown in Figures 156 and 157.

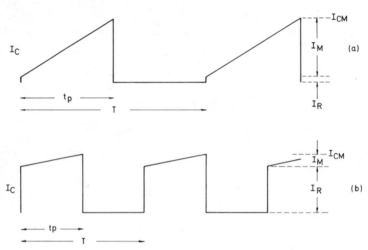

FIGURE 156.   Collector current waveforms of non-saturable transformer inverter: (a) light load; (b) full load.

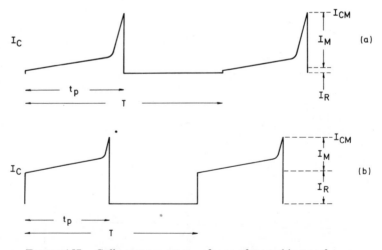

FIGURE 157.   Collector current waveforms of saturable transformer inverter: (a) light load; (b) full load.

Non-saturable Transformer

The collector current waveforms of a non-saturable transformer inverter are shown in Figure 156 with (a) being the waveform for light load and (b) being the waveform for full load. It can be seen that the collector current I_C is a sum of the load current I_R and the magnetising current I_M.

The peak collector current depends on the available base drive, which in the case of the inverter in Figure 155 is constant, as it is determined by the number of turns in the feedback winding and the supply voltage. The peak collector current I_{CM} is thus a product of the transistor gain h_{FE} and the base current I_B.

$$I_{CM} = h_{FE}I_B, \tag{115}$$

which is also equal to

$$I_{CM} = I_R + I_M, \tag{116}$$

where I_R, the collector load current, includes the feedback current required for the oscillation, and the magnetising current I_M depends on the value of the inductance of the primary winding, L_p, and the voltage applied across the primary winding V_p.

$$I_M = \frac{V_p t_p}{L_p}. \tag{117}$$

Substituting for I_M in Equation (116) gives

$$I_{CM} = I_R + \frac{V_p t_p}{L_p}, \tag{118}$$

which indicates that the time of the half-cycle, t_p, must vary if the load current is made to vary, since the remaining terms in the Equation (118) are constant. This is shown in Figure 156(a) and (b), from which it can be seen that, since the half-cycle time t_p varies with the load current I_R, there will be considerable variation of operating frequency with load. This presents a serious disadvantage as it not only affects the efficiency of the circuit but also the filtering and decoupling components. Therefore the inverter with a single non-saturable transformer is not normally used in practice.

Saturable Transformer

Figure 157 shows the collector current waveforms of a saturable trans-former inverter operating with light load in (a) and full load in (b).

It can be seen that the collector current I_C is again made up of a sum of the magnetising current I_M and the load current I_R, so that the same Equations (115) and (116) apply, except that the magnetising current in this case is not a linear function of time, but depends on the properties of the core material used. The transformer is designed to saturate using square loop material, so that the inductive current is initially very small and increases rapidly as saturation is reached. This is shown in Figure 157(a) and (b) which indicates that there is only very small variation in the half-cycle time t_p, so that the operating frequency is made relatively independent of the load current. This compares favourably with the non-saturable transformer circuit, where a considerable variation of frequency with load will be experienced, and the transformer size required to handle the same power will be three to four times as big as one using a saturable transformer.

Inverter With Saturable Transformer

The basic circuit of a push–pull inverter with saturable transformer (Ref. 126) is shown in Figure 158. The circuit is self-oscillating with transistors in common emitter connection.

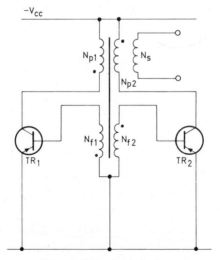

FIGURE 158.   Basic inverter with saturable transformer.

Principle of Operation

Assume the transistor TR_1, of Figure 158, is conducting. The collector current flowing from the battery through the transistor into the dotted end of the primary winding, N_{p1}, makes the polarities of all windings positive at the dotted ends.

The voltage induced across the feedback winding N_{f1}, drives the transistor TR_1 further into conduction, and the voltage induced across N_{f2} keeps the transistor TR_2 cut off. This causes regeneration which rapidly drives the transistor TR_1 into bottoming. At this stage almost all of the battery voltage V_{cc} appears across the collector winding N_{p1}, as a result of which the current and hence the core flux ϕ increase linearly until positive saturation is reached. The collector current then increases rapidly to its peak value of h_{FE} times the base current which is determined by the drive.

Having reached its maximum value, the collector current can no longer increase; thus the induced voltages drop to zero. The collector current in turn starts to fall and induces voltages of the opposite polarity in all windings. This action cuts off the transistor TR_1 and switches on the transistor TR_2.

The supply voltage, switched now across the second half of the primary N_{p2}, causes the flux to decrease linearly to its negative saturation level. At this point, however, the transistor TR_2 comes out of bottoming and the cycle is repeated.

During the switching cycle a peak voltage appears across each transistor as it is cut off. Neglecting the voltage drop across the conducting transistor, say, TR_2, and the transformer winding, the peak inverse voltage across the transistor TR_1 is then the supply voltage plus the voltage induced in the primary winding N_{p1} by the current flowing in the second half of the primary winding N_{p2}. Thus the reverse collector-to-emitter voltage is approximately twice the supply voltage, or $2V_{cc}$.

Starting Circuits

Resistor Starting Circuit

Minimum Base Current Method. The basic push–pull common emitter circuit shown in Figure 158 will not necessarily start to oscillate when the supply is connected as both transistors are cut off and the loop gain is less than one. The transistors, therefore, need to be initially biased into conduction. This is achieved by means of a potentiometer network R_1

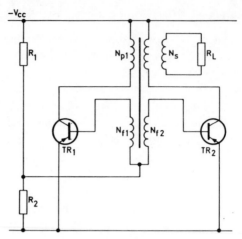

FIGURE 159. Inverter circuit showing
resistor starting.

and R_2 shown in Figure 159. The bias voltage is adjusted so that the loop gain is greater than one and the circuit will always start to oscillate.

The values of the resistors required for starting may be found by considering the loop gain of the stage with a given load (Ref. 128). For that purpose a reference is made to a basic transistor stage shown in Figure 160, where R'_L is the total resistance load appearing across the primary of the transformer, R_b is the external base resistance including the resistance of the feedback winding, n is the feedback turn ratio from the primary winding, and V_{bb} is the voltage which is applied to the base circuit of the transistor.

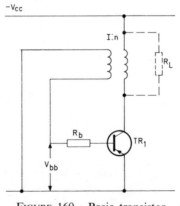

FIGURE 160. Basic transistor
stage.

The voltage gain of the transistor stage shown in Figure 160 is given by

$$\text{voltage gain} = \frac{dV_c}{dV_{bb}}$$

$$= \frac{dV_c}{dI_c} \times \frac{dI_c}{dV_{bb}}$$

$$= R_L' \times G_m, \tag{119}$$

where V_c is the collector voltage, I_c is the collector current, and G_m is the mutual conductance of the stage. The loop gain of the circuit is then

$$\frac{G_m R_L'}{n}. \tag{120}$$

The value of the mutual conductance G_m is given by

$$G_m = g_m \left[\left(1 + \frac{r_{bb'}}{r_{b'c}} \right)^2 + \{ r_{bb'} \omega (C_{b'e} + C_{b'c}) \}^2 \right]^{-\frac{1}{2}} \tag{121}$$

If capacitances $C_{b'e}$ and $C_{b'c}$ are ignored the expression (121) becomes

$$G_m = \frac{g_m}{1 + r_{bb'}/r_{b'c}}, \tag{122}$$

but

$$g_m = \frac{h_{FB}}{r_e} \tag{123}$$

and

$$r_{b'c} = \frac{r_e'}{1 - h_{FB}}; \tag{124}$$

therefore

$$G = \frac{h_{FB}/r_e}{1 + r_{bb'}(1 - h_{FB})/r_e}$$

$$= \frac{h_{FB}}{r_e + r_{bb'}(1 - h_{FB})} \tag{125}$$

or

$$\frac{1}{G_m} = \frac{r_e}{h_{FB}} + r_{bb'} \frac{1 - h_{FB}}{h_{FB}}, \tag{126}$$

which may be approximated to

$$\frac{1}{G_m} = r_e = \frac{r_{bb'}}{h_{FE}}.$$ (127)

Since

$$h_{FB} \approx 1$$

and

$$h_{FE} = \frac{h_{FB}}{1 - h_{FB}}.$$

The $r_{bb'}$ in Equation (127) must include any external base resistance R_b. By substituting therefore $R_{bb} = r_{bb'} + R_b$ and $r_e - 0.025/I_E$, where I_E is the d.c. emitter current in amperes, Equation (127) becomes

$$\frac{1}{G_m} = \frac{0.025}{I_E} + \frac{R_{bb}}{h_{FE}}.$$ (128)

For oscillation the loop gain, Equation (120), should be greater than 1; therefore

$$\frac{G_m R_L'}{n} > 1.$$ (129)

where n is the turns ratio between the primary and the feedback winding;

$$\frac{R_L'}{n} > \frac{1}{G_m}.$$ (130)

Substituting $1/G_m$ from (128), the expression becomes

$$\frac{R_L'}{n} > \frac{0.025}{I_E} + \frac{R_{bb}}{h_{FE}},$$

from which

$$I_E > \frac{0.025}{R_L'/n - R_{bb}/h_{FE}}.$$ (131)

But $I_E = (1 + h_{FE})I_B \approx h_{FE}I_B$; therefore

$$I_B > \frac{n}{40(h_{FE}R_L' - nR_{bb})}.$$ (132)

Equation (132) thus gives the minimum value of I_B per transistor required for oscillation.

For capacitive load, more bias current may be required. This will depend on the value of the capacitor and the leakage inductance of the transformer. To reduce the effect of the capacitive load, in particular in the case of rectifying circuits, it may be necessary to use either a resistive or choke input filter.

The value of R_1 may then be calculated approximately from

$$R_1 = \frac{V_{cc}}{2I_B + (V_{be} + I_B R_{bb})/R_2},\tag{133}$$

where V_{be} is the minimum value of the base–emitter voltage for the required I_B.

The value of R_2 is not critical, and is usually a compromise between the power dissipation due to the oscillatory current and the power drain from the battery due to the bias chain formed by R_1 and R_2.

Minimum Base Voltage Method. In addition to the above-described method for arriving at suitable values of the starting resistors there is an alternative method which is often used.

The method relies on the fact that there is a certain minimum value of the base–emitter voltage V_T required for a transistor to start conducting.

The value of V_T is assumed to be 200 mV for germanium transistors and 600 mV for silicon transistors.

If I_B is the minimum base current to give $I_{IN} = P_{out}/\eta V_{cc}$ where I_{IN} is the supply current, which is approximately equal to the collector current I_C, and V_{BE} is the base–emitter voltage to give I_C, then the feedback voltage required across the base winding N_f is given by

$$V_f = V_{BE} + I_B \frac{R_1 R_2}{R_1 + R_2} - V_T,\tag{134}$$

where R_1 and R_2 are the starting resistors.

In most cases the supply voltage is very much greater than the base voltage V_T, therefore R_1 is much greater than R_2 and the above equation for V_f can be simplified to

$$V_f = V_{BE} + I_B R_2 - V_T.\tag{135}$$

To find the values of the starting resistors, the power dissipation in the input base circuit is considered.

The power delivered by the feedback winding, P_f, is

$$P_f = V_{BE}I_B + I_B{}^2\frac{R_1 R_2}{R_1 + R_2} - V_T I_B \tag{136}$$

and the power drawn from the supply, P_R, by the bias chain is

$$P_R = \frac{V_{cc}{}^2}{R_1 + R_2}. \tag{137}$$

Hence the total power P_F absorbed by the drive circuit is

$$P_F = V_{BE}I_B + I_B{}^2\frac{R_1 R_2}{R_1 + R_2} - V_T I_B + \frac{V_{cc}{}^2}{R_1 + R_2}, \tag{138}$$

but

$$\frac{V_{cc}}{V_T} = \frac{R_1 + R_2}{R_2}; \tag{139}$$

hence

$$R_1 = R_2 \frac{V_{cc} - V_T}{V_T} \tag{140}$$

and

$$R_1 + R_2 = \frac{V_{cc}R_2}{V_T}. \tag{141}$$

Substituting for R_1 and $R_1 + R_2$ in Equation (138) gives

$$P_F = V_{BE}I_B + I_B{}^2 R_2\frac{V_{cc} - V_T}{V_{cc}} - V_T I_B + \frac{V_{cc}V_T}{R_2}. \tag{142}$$

For minimum drive power $dP_F/dR_2 = 0$. Therefore

$$I_B{}^2\frac{V_{cc} - V_T}{V_{cc}} - \frac{V_{cc}V_T}{R_2{}^2} = 0$$

or

$$R_2 = \frac{V_{cc}}{I_B}\left(\frac{V_T}{V_{cc} - V_T}\right)^{\frac{1}{2}} \tag{143}$$

and, since $V_{cc} \gg V_T$,

$$R_2 = \frac{(V_{cc}V_T)^{\frac{1}{2}}}{I_B}. \tag{144}$$

The value of R_1 is found from Equation (140) by substitution of R_2 from Equation (143). Hence

$$R_1 = \frac{V_{cc}}{I_B} \left(\frac{V_{cc} - V_T}{V_T} \right)^{\frac{1}{2}} \tag{145}$$

As before $V_{cc} \gg V_T$; therefore

$$R_1 = \frac{V_{cc}}{I_B} \left(\frac{V_{cc}}{V_T} \right)^{\frac{1}{2}}$$

or

$$R_1 = \frac{V_{cc}}{V_T} \times \frac{(V_{cc} V_T)^{\frac{1}{2}}}{I_B}, \tag{146}$$

where the second term is equal to the value of R_2 in Equation (144); therefore

$$R_1 = R_2 \frac{V_{cc}}{V_T}. \tag{147}$$

Diode Starting Circuit

As shown in Figure 161, the resistor R_2 can be replaced by a diode D_1, which provides a high resistance when starting and a low resistance, with an almost constant voltage, when working normally (Ref. 128).   In this case, lower base circuit dissipation and lower battery drain are achieved, which result in higher overall efficiency.

The value of R_1 for a diode starting circuit can be calculated from

$$R_1 = \frac{V_{cc}}{2I_B + I_D}, \tag{148}$$

where I_B is given by Equation (132) and I_D is the reverse current of the diode which for silicon diodes is normally small enough to be neglected.

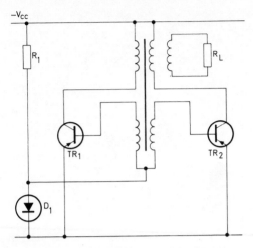

FIGURE 161.   Inverter circuit with diode
starting.

Summary of Starting Circuits

Many different starting circuits have been developed for use with inverters and converters. The most often-used arrangements are derived from the basic amplifier networks.

The common emitter starting circuits are shown in Figure 162, and common collector starting circuits are shown in Figure 163.

Design of a Saturable Transformer

The design of the transformer is unconventional since it saturates during part of a cycle. Also, for fast switching action, close coupling between primary and feedback winding is required. In addition the two halves of the primary should be bifilar-wound to achieve good balance and maximum coupling between the two sections. This reduces leakage inductance and practically eliminates the high peak voltages which otherwise might appear across the cut off transistors and cause damage due to avalanche breakdown.

When a d.c. inverter or converter is required for a particular application, the input voltage, output voltage, and the output power will usually be known. It is therefore necessary to choose the frequency of the operation and to calculate the required number of primary turns. Other windings are related to the primary by normal turns ratios and the voltages required.

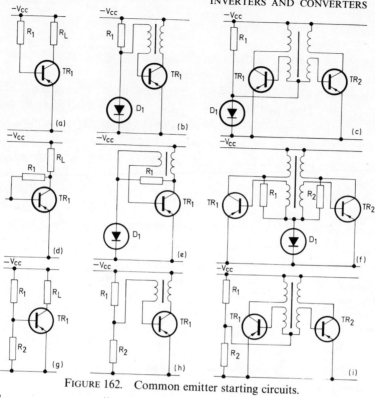

FIGURE 162. Common emitter starting circuits.

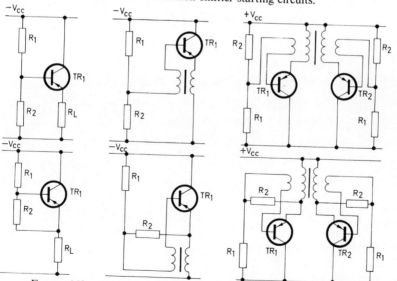

FIGURE 163. Common collector (emitter follower) starting circuits.

Operating Frequency

The choice of operating frequency is primarily governed by the efficiency expected and the size of the converter.

In general, transformer core losses and transistor transient losses increase with frequency (Figure 128); therefore the frequency of operation should be made as low as possible. On the other hand, the physical size of the transformer decreases as the operating frequency is increased. These problems will be further discussed in conjunction with the high-frequency inverters.

An expression for the frequency of the oscillation may be derived by considering a voltage applied across an inductance which in this case is the primary winding N_{p1} or N_{p2}, of Figure 159, of the transformer.

The voltage developed across the primary winding V_p, is approximately equal to the supply voltage V_{cc} less the saturation voltage of the transistor $V_{CE(sat)}$ and the voltage drop across the resistance of the primary winding V_{R_p}. Therefore

$$V_p = V_{cc} - V_{CE(sat)} - V_{R_p}. \tag{149}$$

The voltage V_p is also given by

$$V_p = N_p \frac{d\phi}{dt} \times 10^{-8}, \tag{150}$$

where $N_p = N_{p1} = N_{p2}$, is the number of primary turns and ϕ is the flux in the transformer core. If t_p is the time of one half-cycle of oscillation, during which the saturation flux ϕ undergoes complete reversal, then

$$\int_0^{t_p} V_p \, dt = \int_{-\phi_s}^{+\phi_s} -N_p \, d\phi \times 10^{-8}$$

or

$$V_p t_p = 2N_p \phi_s \times 10^{-8}. \tag{151}$$

But

$$\phi_s = A \times B_s, \tag{152}$$

where B_s is the saturation flux density in gauss and A is the cross-sectional area in square centimetres, so that

$$t_p = \frac{2N_p A B_s \times 10^{-8}}{V_p}; \tag{153}$$

hence the frequency of the oscillation is

$$f = \frac{1}{2t_p} = \frac{V_p 10^8}{4N_p A B_s}.$$ (154)

Core Material

For the frequency to be largely independent of the load current, the collector current rise must be rapid as soon as the transformer core starts to saturate. This is illustrated in Figure 157 and suggests a core material such as HCR alloy. The material has a square hysteresis loop, a high saturation flux density, and a low hysteresis loss.

For best efficiency the optimum operating frequency will generally lie between 200 and 600 Hz. At higher frequencies core losses increase rapidly so that material other than HCR may be required.

Primary Winding

Having decided on the frequency of the operation and the type of core material, another condition that must be satisfied for correct operation is that there should be sufficient current available to saturate the core. This takes into account the number of turns on the transformer primary winding and the length of magnetic path, l. The relation is given by a well-known transformer equation:

$$H = \frac{4\pi N_p I_L}{10l} \text{ or } \frac{1 \cdot 26 N_p I_L}{l},$$ (155)

from which for saturation

$$H_s = \frac{1 \cdot 26 N_p I_L}{l} > H_0,$$ (156)

where H_s is the value of the magnetising field strength at saturation, expressed in oersteds, H_0 is the intrinsic strength of the magnetising field of the material used for the core, expressed in oersteds, I_L is the value of the inductive current in amperes, and l is the length of the magnetic path in centimetres.

In Equation (156), H_s is fixed by the material chosen for the core, and l by the size of the laminations. Therefore a value for N_p can be obtained for a particular value of the inductive current I_L, so that

$$N_p = \frac{H_s l}{1 \cdot 26 I_L},\tag{157}$$

This value of N_p can now be substituted into Equation (154) in order to establish the necessary cross-sectional area for the required operating frequency.

$$\text{Cross-sectional area } A = LTS,\tag{158}$$

where L is the lamination width, in centimetres, T is the lamination thickness in centimetres, and S is the number of laminations.

Feedback Winding

The feedback windings must be designed to ensure that adequate current is available from the transistors at the minimum supply voltage. In addition the design must ensure that, with the supply voltage at its maximum, the peak current in the transistor does not exceed the maximum rated current. This involves the spreads in the transistor characteristics and accounts for the use of the external base resistor mentioned in deriving the minimum value of the base current required for reliable starting (Equation (132)). The number of turns required for the feedback winding, N_f, is given by

$$N_f = N_p \frac{V_p}{V_f},\tag{159}$$

where V_f, the required feedback voltage, is given by

$$V_f = V_{BE} + I_B R_b + V_{D1},\tag{160}$$

where V_{BE} is the base–emitter voltage for the peak collector current, I_B the base current required for the peak collector current, R_b the external base resistor, and V_{D1} the forward voltage drop across the starting diode D_1.

The external base resistor is used to compensate for the individual differences between the V_{BE} characteristics of the transistors. In order to provide sufficient balancing action, the voltage drop across R_b should be greater than the maximum V_{BE} of the transistor used, so that

$$R_b \geqslant \frac{V_{BEM}}{I_B}$$

In cases where excessive base circuit dissipation results, because of R_b, the resistor can be omitted if slight unbalance due to differences in V_{BE} of the transistors can be tolerated.

The resistor R_b also has the undesirable effect of increasing the hole storage time of the transistor and thus increasing transient losses during switch-off.

In order to reduce the hole storage time, a capacitor can be connected across each base resistor R_b.

Secondary Winding

The number of the secondary turns, N_s, depends on the output voltage required and the type of rectifying circuit used, and can be found from

$$V_{out} = V_p \frac{N_s}{N_p} - R_{out} I_{out} , \tag{161}$$

where V_{out} is the output voltage, R_{out} the output impedance of the converter including the rectifying circuit, and I_{out} is the output current. The second term in Equation (161), representing the regulation, is small in general, and may therefore often be neglected in calculating the number of turns required. Therefore

$$N_s = N_p \frac{V_{out}}{V_p} . \tag{162}$$

General Remarks

The push–pull inverter circuit provides an almost constant-voltage output characteristic, the power drawn being dependent on the loading. The power loss remains approximately constant as the loading varies and efficiency therefore increases as the power output increases. Above a certain value of collector current, the transistor gain falls, so that driving losses increase as well as transient and resistive losses and the efficiency decreases.

Most push–pull arrangements also make use of the ability of transistors to pass pulses of current in the reverse direction. By this means the transistors return the stored energy in the transformer to the battery if the loading is very light. This usually happens when the load current, reflected to the primary, is less than the magnetising current. The energy stored owing to the excess of the magnetising current is returned to the battery through the transistor that has just been switched on, but is conducting in the reverse direction until the magnetising current reverses and the collector current starts flowing in the forward direction.

There are three principal disadvantages associated with the conventional push-pull d.c. inverter circuit described above. These disadvantages are summarised as follows.

(i)   The peak collector current is determined by the available drive and is independent of the load; therefore high peak currents exist even with light loads. Also, for a given power delivered to the load, the ratio of peak current to load current is high.

(ii)   The design is affected by spreads in transistor characteristics unless the transistors are selected, or each circuit is individually adjusted. The peak collector current is determined by h_{FE} and V_{be}. Therefore, for the circuit to operate with full transistor spreads, it must be designed so that the peak current rating of the transistors is not exceeded when they have a high h_{FE} and low V_{be}. If transistors with low h_{FE} or high V_{be} values are used in this circuit, the peak collector current will be much lower than the rated value, so that the maximum usable load current and thus the available output power are reduced considerably. The use of an external base resistor reduces the unbalance arising from spreads in V_{be}, but, even then, the collector load current is limited to about a third of the rated peak current, because of the spreads in h_{FE}.

(iii)   The transformer uses an expensive core material which has a square hysteresis loop and a high value of flux density at saturation.

These disadvantages can be overcome using two transformers in a circuit such as that described in the following section.

Inverter With Two Transformers

The two-transformer circuit differs from the conventional type of push–pull inverter with single saturable transformer in that a small saturable drive transformer is used to control the switching and a larger transformer, working linearly, steps up the output to the required value. The essential improvement is that a higher proportion of the transistor peak current rating can be used with all transistors and, therefore, the output power can be increased while tolerating full production spreads in the characteristics of the transistors.

In addition, the overall cost of the inverter is reduced owing to the fact that only the small drive transformer uses an expensive core material. The output transformer is of conventional type with a cheap silicon iron core.

A basic two-transformer inverter circuit is shown in Figure 164 (Refs. 127 to 129).

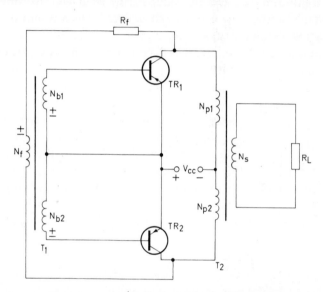

FIGURE 164. Basic two-transformer d.c. inverter circuit.

Circuit Operation

On connecting the supplies to the circuit in Figure 164, because of unbalance in the circuit, one of the transistors, say, TR_1, will conduct, causing its collector voltage to swing (to zero) by very nearly the supply voltage. The voltage, building up across the primary of the output transformer is applied across the primary of the drive transformer T_1, in series with a feedback resistor R_f. The secondary windings are so arranged that the transistor TR_2 will be reverse-biased and will remain cut off, and TR_1 will be held in the bottomed condition.

As soon as the core of transformer T_1 reaches saturation, rapidly increasing primary current causes an additional drop across the feedback resistor R_f. This drop reduces the drive, and the collector current of transistor TR_1, which was bottomed, starts to decrease, causing in turn the reversal of the polarities of the voltages in all windings. Transistor TR_1 is rapidly driven to a cut-off state and transistor TR_2 is switched on. This transistor continues in this state until negative saturation of the transformer is reached. The voltage switches back to the initial state and the cycle is repeated. The oscillation then continues at a frequency determined by the design of the saturable transformer T_1 and the value of the feedback resistor.

For reliable starting the transistors are initially biased into conduction by using a resistor and a diode (R_1 and D_1, Figure 165). The external base resistors are added to reduce the effect of spreads in V_{BE} on the operation of the circuit.

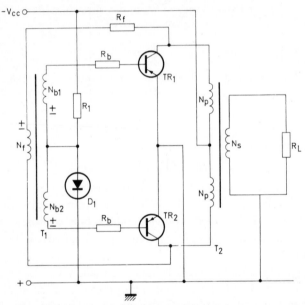

FIGURE 165. Basic circuit with starting arrangement
added.

The collector current in either of the transistors rises to the effective sum of the load current, the magnetising current of the output transformer, and the feedback current needed to produce the drive. Because the output transformer is not allowed to saturate, the magnetising current is only a small fraction of the load current.

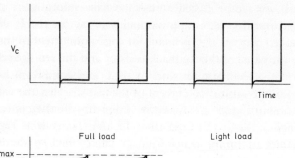

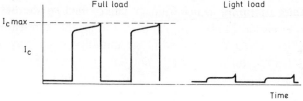

FIGURE 166. Collector voltage and current waveforms
for a purely resistive load.

The collector voltage and current waveforms, for a purely resistive
load, are shown in Figure 166.

Design Considerations

The design of an inverter is normally based on the available supply
voltage, the required output voltage, and output power.

The peak voltage, at the collector of either transistor when cut off, is
approximately twice the supply voltage. The supply voltage should always
be less than half of the collector breakdown voltage at the peak value of
inductive current.

The design of the transformers is not critical and a wide choice of operating frequency can be tolerated, depending on the required size and weight
of the inverter and its efficiency.

Drive Transformer

The primary of the drive transformer is connected in series with a feedback
resistor R_f across the two collectors of the transistors of the converter
(Figure 165).

The peak voltage produced across the two collectors by the primaries of the output transformer is approximately twice the supply voltage. The voltage applied across the primary of the input transformer, however, depends on the value of the feedback resistor and the required drive current. The value of the feedback resistor, in turn, is a compromise between the requirement of the saturation current of the transformer, the voltage applied across the primary, and the operating frequency of the converter.

The design of the drive transformer T_1 follows the same lines as that for the inverter with a saturable transformer. Therefore the operating frequency is given by Equation (154):

$$f = \frac{V_f 10^8}{4N_f AB_s}.$$
(163)

V_f is the voltage applied across the primary winding of transformer T_1, where V_f replaces V_p and N_f replaces N_p. The number of primary turns, N_f, is given by (see Equation (157)).

$$N_f = \frac{H_s l_c}{1 \cdot 26 I_L},$$
(164)

and the number of turns required for the base windings N_b is given by

$$N_b = N_f \frac{V_f}{V_{bb}},$$
(165)

where

$$V_{bb} = V_{BE} + I_B R_B + V_{D1}.$$
(166)

The value of the feedback resistor R_f is then calculated from

$$R_f = \frac{2V_{cc} - V_f}{I_f},$$
(167)

where I_f is the current in the primary of the feedback transformer T_1, and is the sum of (i) the base current required to bottom the transistor at full load multiplied by N_b/N_f and (ii) the magnetising current, I_L in Equation (157), just before saturation commences, that is I_m, the linear magnetising part of I_L. Therefore

$$I_f = I_B \frac{N_b}{N_f} + I_m.$$
(168)

Output Transformer

The output transformer T_2 is a normal linear transformer and is designed using conventional techniques. The primary windings must have a sufficiently high inductance to keep the required value of magnetising current low. Also the leakage inductance must be negligible, which is possible if bifilar windings are used.

The value of inductance required for each half of the primary can be calculated from

$$L_p = V_p \frac{t_p}{I_M},\tag{169}$$

where V_p is the voltage across the primary winding of the conducting transistor and is given by

$$V_p = V_{cc} - V_{CE(sat)} - V_{Rp},\tag{170}$$

where t_p is the time of half a cycle, I_M the magnetising current of the transformer T_2, V_{cc} the supply voltage, $V_{CE(sat)}$ the saturation voltage of the transistor, and V_{Rp} is the voltage drop across the resistance of the primary winding of T_2. Since $t_p = 1/2f$, therefore

$$L_p = \frac{V_p 10^3}{2f \times I_M} \text{ mH.}\tag{171}$$

With transformer core materials such as Ferroxcube there is no problem of obtaining the number of turns required for the primary winding. The manufacturers of the core material specify turns required for one millihenry in a form of a constant α. Therefore the number of turns required for the primary winding is

$$N_p = \alpha \, (L_p)^{\frac{1}{2}}\tag{172}$$

where L_p is the inductance in millihenries.

In the case of laminated iron cores and strip-wound C cores, insufficient information is available.

To ensure that the transformer is operating at a correct value of magnetising current, and at the same time has the required inductance, it is necessary to satisfy the following two equations for N_p:

$$N_p = \frac{10H(l_c + \mu l_g)}{4\pi I_M} \tag{173}$$

and

$$N_p = \left\{\frac{L_p(l_g + l_c/\mu)10^6}{4\pi A}\right\}^{\frac{1}{2}}, \tag{174}$$

where μ is the permeability of the core material, l_c the length of flux path in centimetres, l_g the length of air gap in centimetres, and L_p the inductance in millihenries.

The number of secondary turns is approximately given by

$$N_s = N_p \frac{V_s}{V_p}, \tag{175}$$

and the maximum available power from the inverter is given by

$$P_{out} = \eta V_{cc} I_{av}, \tag{176}$$

where η is the expected efficiency, and I_{av} is the average current from the supply which is approximately equal to the collector load current I_R, where

$$I_R = I_{CM} - I_M - I_f, \tag{177}$$

so that

$$P_{out} = \eta V_{cc} I_R. \tag{178}$$

Starting Circuit

The circuit shown in Figure 164 will not necessarily start to oscillate, especially when heavily loaded, because both transistors are initially cut off. A permanent bias is therefore applied so that the circuit has a loop gain greater than unity and will always start to oscillate.

To ensure a loop gain greater than unity, the base current I_b must be greater than

$$\frac{n}{40(h_{FE}R'_L - nR_{bb})} \tag{179}$$

where R_L' is the resistive load appearing across the primary winding of T_2, R_{bb} is the total base resistance, both internal and external, including R_f, and n is the turns ratio of the feedback winding, allowing for both transformers.

The value of R_{bb} for the two-transformer circuits with feedback resistor R_f is given by

$$R_{bb} = R_b + r_{bb'} + R_f\left(\frac{N_b}{N_f}\right)^2. \tag{180}$$

Feedback Resistor

The optimum value of the feedback resistor R_f is found to be that value which will drop about half the available voltage, at the drive current corresponding to the maximum load current.

Increasing R_f causes a greater drop in voltage across it, so that less voltage is available across the primary of the drive transformer. As inferred from Equation (163), the operating frequency will decrease.

Decreasing R_f will increase the operating frequency and increase the losses arising from the circuit resistance and in the transformer core, owing to a higher magnetising current.

Design of a Practical Circuit

Several factors affect the design of a practical two-transformer d.c. converter; these factors and the performance of the circuit are now examined in detail.

Operating Frequency

The choice of operating frequency is not very critical and will depend on the physical size and the efficiency of the converter.

Although losses in the transformer cores and transient losses of the transistors increase with operating frequency, the efficiency varies only a few per cent over the frequency range 300 to 1000 Hz.

Drive Transformer

Since the required drive is less than one watt, only a small core is needed.

A square stack of Telcon HCR alloy laminations (pattern 224) can be used. This material has the following characteristics:

$$B_s = 15\,000 \text{ G}, \qquad H_s = 2 \text{ Oe}, \qquad l = 5 \cdot 72 \text{ cm}.$$

The cross-sectional area (A) of 50 laminations equals $0 \cdot 331$ cm^2.

If the magnetising current I_L is 40 mA, then, by substitution in Equation (164),

$$N_f = \frac{2 \times 5 \cdot 72}{1 \cdot 26 \times 40 \times 10^{-3}}.$$

Thus there should be 227 turns on the feedback winding of the drive transformer.

To evaluate Equation (163), the value of the feedback voltage V_f and of the base winding voltage V_{bb} must first be calculated.

The maximum base-to-emitter voltage $V_{be(max)}$, required for the lower limit of OC28 transistors, is $1 \cdot 6$ V and the maximum base current $I_{B(max)}$ is 375 mA for the maximum peak collector current of 6 A. With an external base resistor of 10 Ω and allowing 1 V across the starting diode, at 375 mA transistor base current the required feedback voltage (calculated from Equation (166) is $6 \cdot 35$ V. If a turns ratio of 4 : 1 is chosen, 57 turns are required for each feedback winding and the primary current I_f is about 94 mA. The voltage developed across the primary under these conditions is

$$\begin{aligned} V_f &= n \times V_{bb} \\ &= 4 \times 6 \cdot 35 \\ &= 25 \cdot 4 \text{ V}. \end{aligned}$$

From Equation (163) the frequency now equals

$$f = \frac{25 \cdot 4 \times 10^8}{4 \times 227 \times 0 \cdot 331 \times 15 \times 10^3}$$

$$= 564 \text{ Hz}.$$

The value of the feedback resistor R_f is given by Equation (167):

$$R_f = \frac{(2 \times 28) - 25 \cdot 4}{94 \times 10^{-3}}$$

$$= 326 \ \Omega.$$

The nearest preferred value of 330 Ω should be used in a practical circuit.

Transistor Spreads

The drive transformer, designed in the previous section, is intended to drive a circuit containing the worst possible transistors.

The performance of the converter, with transistors having high h_{FE} and low V_{BE} will be modified very little except that the operating frequency might decrease by a maximum of 14% from the value calculated. The frequency may be adjusted, if required, by extracting a few laminations from the core of the drive transformer. With the components shown in the circuit in Figure 167 and typical transistors, the operating frequency is about 500 Hz.

For maximum spreads in transistor characteristics, the change in output voltage, output power, and efficiency will be less than 3%.

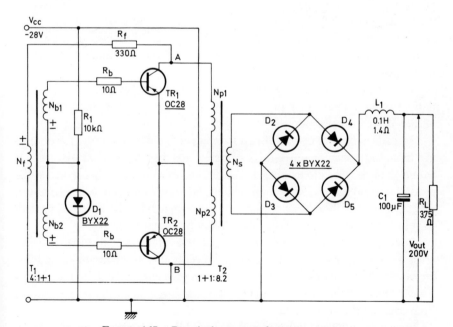

FIGURE 167. Practical two-transformer converter.

Output Transformer

If the magnetising current of the output transformer is to rise to 400 mA during the half-cycle time t (equals 1 μs), the value of inductance required for each half of the primary, as given by Equation (169), is

$$L_p = 28 \times \frac{1 \times 10^{-6}}{400 \times 10^{-3}}$$

$$= 700 \text{ mH.}$$

To avoid excessive loss of power, the resistance of each primary winding should be less than 0·2 Ω.

The peak collector current is the sum of the magnetising current, the the circuit can operate up to 5·5 A load current, provided the two halves of the circuit and the transistors are identical. In practice, owing to slight unbalance in the circuit and to the fact that the transistors are not matched, the out-of-balance current through the output transformer causes some pre-magnetisation of the core. As a result, the collector current of one of the transistors will rise to a higher peak value than the other. If the circuit has been designed for operation up to the maximum ratings of the transistors, the peak collector current can thus exceed the safe value if the circuit is not modified.

The unbalance of the circuit can be reduced by using bifilar windings, both for the primaries of the output transformer and for the secondaries of the drive transformer.

Spreads in h_{FE} are more difficult to deal with; the best method for obtaining balanced collector currents is to use a matched pair of transistors. It would then be possible to operate the circuit up to the full theoretical value of load current, 5·5 A, with a consequent increase in output power of about 20%. External base resistors can be used to reduce the effect of spreads in V_{BE}.

If no precautions are taken to avoid the unbalance, the load current must be limited to 4·5 A, and the peak collector current must not exceed 5 A including the feedback current and the magnetising current of the output transformer. This allows for up to 1 A of out-of-balance current plus surges from the smoothing system.

Even within these limitations it is possible to obtain an output power of 100 W with a 28 V supply and with the additional advantage of using transistors with full spreads.

With matched transistors and a purely resistive load, 130 W output can be obtained at about 90% efficiency.

Starting Circuit

On full load, the reflected load resistance is approximately 5·6 Ω. With a feedback turns ratio of 2 (see page 244), a minimum low-current h_{FE} of 20, and R_{bb} equalling 35·6 Ω, the minimum base current required for oscillation is (see Equation (179))

$$I_b > \frac{2}{40(20 \times 5 \cdot 6 - 2 \times 35 \cdot 6)}$$

$$> 1 \cdot 23 \text{ mA.}$$

Diode Starting. Using a diode to initiate oscillation, the value of R_1 is found from Equation (148)

$$R_1 = \frac{28}{2 \times 1 \cdot 23 + 10^{-3} + 0 \cdot 02 \times 10^{-3}}$$

$$= 11 \cdot 3 \text{ k}\Omega.$$

Resistor Starting. The value of R_1 for resistor starting (Equation (133)), when R_2 is 3·3 Ω and the maximum required V_{be} equals 300 mV, is

$$R_1 = \frac{28}{2 \times 1 \cdot 23 \times 10^{-3} + (300 \times 10^{-3} + 1 \cdot 23 \times 10^{-3} \times 35 \cdot 6)/3 \cdot 3}$$

$$= 263 \ \Omega.$$

These values for R_1 and R_2 would considerably lower the efficiency of the circuit. The value of R_1 used in the practical circuit was 3·3 kΩ, and starting was found to be satisfactory for both choke input and purely resistive loads. Higher values for R_1 might prevent starting with large capacitive loads.

Heatsink Design

The design of heatsinks is fully described in Chapter 2.   The heatsinks for the practical circuit (Figure 167) were made of copper 3·2 mm thick and of area 180 cm² (suitably folded).   The surface was blackened to assist cooling by radiation.

The thermal resistance of the heatsink, R_{th}, was about 2 degC/W.   The maximum ambient temperature at which the d.c. converter will operate satisfactorily is given by

$$T_{amb(max)} = T_{j(max)} - P_c(R_{th(h)} + R_{th(m)}),$$

where $T_{j(max)}$ is the maximum junction temperature (90 °C), P_c is the collector dissipation, and $R_{th(m)}$ is the temperature rise of the junction above that of the mounting base (1·5 degC/W maximum).   Therefore, for a dissipation of 3 W per transistor,

$$T_{amb(max)} = 90 - 3(2 + 1·5)$$
$$\approx 80 \ °C.$$

Performance

The circuit shown in Figure 167 is that of a practical converter, constructed to a design based upon the procedure detailed previously.   The performance of this converter was thoroughly investigated; details and comments now follow:

supply voltage	28 V,	output voltage	195 V,
supply current	4·3 A,	output current	529 mA,
input power	120 W,	output power	103 W,
frequency	500 Hz,	efficiency	86%,
ripple voltage	220 mV.		

Over the range of temperature from − 10 °C to +80 °C the performance was hardly affected.   Reduction of copper losses in the output transformer can lead to a higher output and an efficiency of about 90%.

The load current and supply voltage were varied over a wide range of values and the results obtained are shown graphically in Figures 168 and 169.

Figure 168 shows the variation with load current of output voltage, output power, efficiency, and operating frequency (curves (a), (b), (c), (d) respectively).   Figure 169 shows the variation of the same quantities with supply voltage (curves (a), (b), (c), (d) respectively).

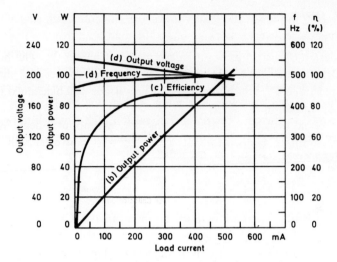

FIGURE 168.   Effect on operation of varying the load current.

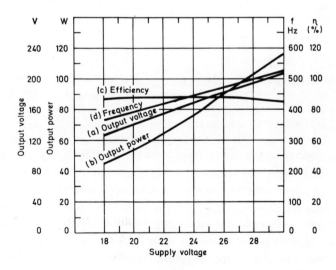

FIGURE 169.   Effect on operation of varying the supply voltage.

Figure 170 shows the collector voltage and current waveforms of the two-transformer d.c. converter working under full load conditions. The collector current waveform for a purely resistive load is shown in Figure 170(b). With a small capacitance across the load, the output transformer starts to ring. As a result the collector current (Figure 170(c)) rises to a higher peak value. If the capacitance is much higher, the oscillation is damped and the collector current does not rise to such a high value (Figure 170(d)). It must be emphasised, at this point, that the maximum peak current rating of the transistors must not be exceeded.

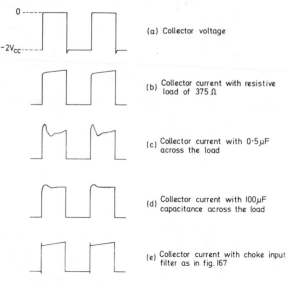

FIGURE 170. Collector voltage and current waveforms for various load conditions.

The disadvantage of a large capacitive load is that it can affect starting when it short-circuits the load initially. If this occurs a surge-limiting resistor could be connected in series with the load, being progressively short-circuited when the converter is switched on.

So that the converter can operate satisfactorily with a large capacitance across the output, it is necessary either to reduce the load current or, much the better solution, to use a resistive or a choke input filter. The collector current waveform when the latter is used is shown in Figure 170(e). The spikes at the beginning of the waveform are due to the inductance of the transformer and choke, and must not exceed the maximum peak current rating.

Satisfactory operation of the d.c. converter (Figure 167) with resistor starting was obtained by reducing the value of R_1 to 3·3 kΩ and replacing the diode D_1 by a resistor R_2, of 3·3 Ω. Performance figures for this modified circuit are given:

supply voltage	28 V,	output voltage	193 V,
supply current	4·36 A,	output current	526 mA,
input power	122 W,	output power	101 W,
frequency	510 Hz,	efficiency	83%.

Summary of Transformer Details

Saturating Drive Transformer T_1.

Core material:	HCR alloy (Telegraph Construction and Maintenance Co. Ltd.), pattern 224, 50 laminations.
Bobbin:	Insulated Components and Materials Ltd., Type 187A.
Primary winding:	227 turns of 34 s.w.g. enamelled copper wire.
Secondary winding:	57 + 57 turns (bifilar winding) of 30 s.w.g. enamelled copper wire.

Linear Output Transformer T_2.

Primary winding:	inductance = 70 mH/winding, resistance < 0·2 Ω/winding, (bifilar winding).
Secondary winding:	resistance < 15 Ω.
Turns ratio:	1 + 1 : 8·2.

Common Collector Connection

A modified two-transformer d.c. converter circuit (Ref. 128) is shown in Figure 171. Although the collectors can be connected to the same heatsink, or directly to the chassis in equipment having the negative side earthed, the circuit operates as a push–pull common emitter amplifier with the input applied between base and emitter. Figure 171 is a redrawn version of Figure 167, with a resistor R_2 in place of diode D_1. The main

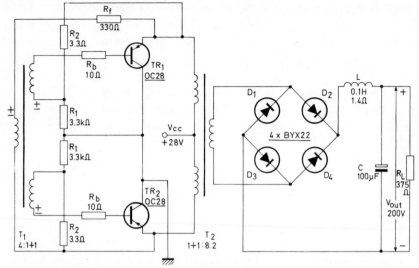

FIGURE 171. This circuit is a further modification of the basic circuit
shown in Figure 167. The emitter and collector connections are reversed
and separate starting resistors are provided for each transistor.

difference between the two circuits is that the collector and emitter connections are interchanged; a separate starting circuit is used for each transistor. Note that in this arrangement a diode must not be used in place
of R_2, because there would be no means of diverting base current and the
transistor would never be cut off.

The performance of this modified circuit is almost identical with that
of the circuit in Figure 167, with resistive starting, except that the efficiency
is one or two per cent lower because of the additional current drain arising
from the separate biasing arrangements. The performance is as follows:

supply voltage	28 V,	output voltage	193 V,
supply current	4·4 A,	output current	522 mA,
input power	123 W,	output power	100 W,
frequency	514 Hz,	efficiency	82%.

Voltage Doubler Output

A voltage doubler arrangement is often required instead of a bridge
rectifier circuit (Ref. 128). A suitable circuit is given is Figure 172. Results
of measurements carried out on this circuit are displayed graphically

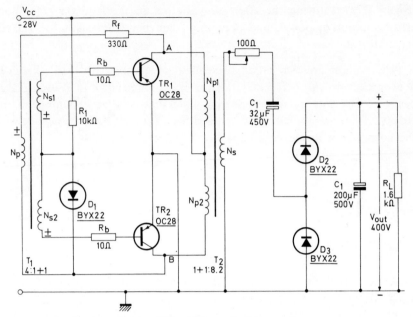

FIGURE 172.   Voltage doubler output.

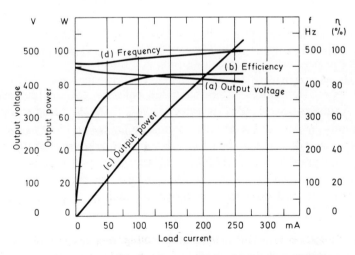

FIGURE 173.   Effect, on the operation of the voltage doubler
circuit of varying the load current.

(Figure 173), showing the effect of variation of the load on regulation, efficiency, output power, and operating frequency (curves (a), (b), (c), (d) respectively).

The voltage doubler circuit presents a large capacitive load; therefore a progressively short-circuited resistor is recommended to prevent large peaks of charging current appearing immediately after switching on the supply. This resistor can be in series either with the supply voltage or with the output transformer and capacitor C_1 (see Figure 172).

Other Examples

A 300 W 400 Hz Inverter

A push–pull two-transformer d.c. inverter circuit has been designed using Mullard ADZ12 transistors for operation at 400 Hz from a 28 V supply (Ref. 135). The output voltage is basically a square wave. If a sine-wave output is required, a suitable filter can be incorporated.

Practical Design Considerations. The transistor has a number of limitations which must be taken into account. First is the maximum collector voltage that may be applied, which in turn limits the maximum allowable supply voltage. Since ADZ12 transistors are rated at $V_{CEM} = -60$ V, the maximum supply voltage is limited to 28 V, allowing for a swing of 56 V across the transistor which is cut off, and leaving sufficient safety margin for small fluctuations.

The second limitation is the maximum collector current that may be used. Although the ADZ12 is rated at $I_{CM} = 20$ A, for d.c. inverter applications the peak collector current must be limited to 15 A, as indicated in the transistor data under the permissible area of operation for the device. The maximum power available from the push–pull pair is given by $P_{max} = \eta V_{cc} I_{CM}$, which is about 300 W allowing for the efficiency of the circuit, the voltage drop across the transistor, and the magnetising current of the transformer.

As indicated on page 210 the available voltage is equally divided between the voltage drop across the feedback resistor R_f and the voltage V_f developed across the primary of the saturating drive transformer T_1 (Figure 174).

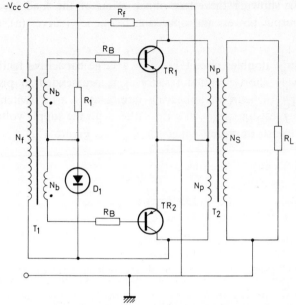

FIGURE 174.   Practical 400 Hz inverter.

The equations listed below are used in the design of transformer T_1:

$$I_B = \frac{I_{CM}}{\beta},$$

$$V_{bb} = V_{BE} + I_B \times R_B + V_{D1},$$

$$n = \frac{V_f}{V_{bb}},$$

$$I_f = \frac{I_b}{n},$$

$$R_f = \frac{2V_{cc} - V_f}{I_f} = \frac{V_{cc}}{I_f}, \text{ since } V_f = V_{cc},$$

$$N_f = \frac{H_s l_c}{1 \cdot 26 I_M},$$

$$N_b = \frac{N_f}{n},$$

$$f = \frac{V \times 10^8}{4 N_f A B_s},$$

where $A = TLS$ and T is the lamination thickness, L the lamination width, and S the number of laminations.

The design values, with a square stack of Telcon HCR alloy laminations, pattern number 225N, are given below:

$$V_{cc} = 28 \text{ V}, \qquad H_s = 2 \text{ Oe},$$
$$V_{BE} = 0.75 \text{ V}, \qquad T = 0.004 \times 2.54 \text{ cm},$$
$$V_{D_1} = 1 \text{ V}, \qquad L = 0.375 \times 2.54 \text{ cm},$$
$$I_{CM} = 15 \text{ A}, \qquad S = 80,$$
$$I_M = 100 \text{ mA}, \qquad l_c = 8.58 \text{ cm},$$
$$B_s = 15\,000 \text{ G}, \qquad R_B = 5 \text{ }\Omega,$$

where l_c is the length of the flux path.

The resulting design is

$$N_f = 136 \text{ turns},$$
$$N_b = 33 \text{ turns},$$
$$R_f = 116.1 \text{ }\Omega \text{ (nearest preferred value} = 120 \text{ }\Omega),$$
$$f = 443 \text{ Hz}.$$

The above results are for a pair of transistors with a minimum current gain h_{FE} of 15.

The output transformer is designed with double-loop C cores, Reference HWR 50/18/5.

A compromise is again necessary, to keep the size of the transformer down. The magnetising current is therefore allowed to rise to 1·5 A during each half-cycle.

Using Equations (169) to (180) and the following data:

$$V_{cc} = 28 \text{ V}, \qquad \eta = 0.8,$$
$$V_c = 1 \text{ V}, \qquad f = 400 \text{ Hz},$$
$$V_s = 230 \text{ V}, \qquad A = 16.9 \text{ cm}^2,$$
$$I_{CM} = 15 \text{ A}, \qquad l_c = 20.625 \text{ cm},$$
$$I_M = 1.5 \text{ A}, \qquad l_g = 0.083 \text{ mm}.$$

the resulting design is

$$P_{out} = 302.4 \text{ W},$$
$$L_p = 22.5 \text{ mH},$$
$$N_p = 50 \text{ turns},$$
$$N_s = 426 \text{ turns}.$$

Performance of the inverter. A practical circuit was constructed on the basis of the above calculations. The measurements carried out on the circuit are given in Table 8.

Table 8. Comparison of high-voltage and low-voltage inverters: two-transformer configuration

Components		High-voltage	Low-voltage	
Transistors		ADZ12	ADZ11	
Starting diode	D_1	BYX38	BYX38	
Starting resistor	R_1	820	680	Ω
Feedback resistor	R_f	120	33	Ω
Performance				
Supply voltage	V_{cc}	28	14	V
Supply current	I	13·4	12·5	A
Input power	P_{in}	375	175	W
Peak base current	I_{BM}	1·3	1·3	A
Peak collector current	I_{CM}	14·8	14·3	A
Output voltage	V_{out}	221	100	V
Load resistance	R_L	150	75	Ω
Output power	P_{out}	327	133	W
Efficiency	$\eta \times 100$	87	76	%
Frequency	f	408	416	Hz
Saturating drive transformer T_1				
Core		HCR alloy pattern 225N+	HCR alloy pattern 225N+	
No. of laminations		80	80	
Feedback winding	N_f	136 27 s.w.g. Lewmex copper wire	68 turns 22 s.w.g. Lewmex copper wire	
Base winding	N_b	33+33 bifilar 22 s.w.g. Lewmex copper wire	33+33 turns bifilar 22 s.w.g. Lewmex copper wire	
Linear output transformer T_2				
Core		Double-loop C core type HWR 50/18/5 ‡	Double-loop C core type HWR 50/18/5 ‡	
Primary winding	N_p	50+50 bifilar 15 s.w.g. Lewmex copper wire	50+50 turns bifilar 15 s.w.g. Lewmex copper wire	
Secondary winding	N_s	426 20 s.w.g. Lewmex copper wire	426 turns 20 s.w.g. Lewmex copper wire	

† Telegraph Construction and Maintenance Co. Ltd.
‡ Telcon-Magnetic Cores Ltd. or English Electric Co. Ltd.

It is worth noting that a higher power is obtained than the value calculated. The discrepancy is due to the fact that for calculation purposes the efficiency was assumed to be 80%, whereas the measured efficiency is 87%. The calculations were repeated after substituting the measured value of efficiency of 87%, and the new calculated value of output power was found to be 328·9 W.

Modification for 14 *V Supplies.*   The circuit was modified for use with 14 V supplies and ADZ11 transistors which have the same current rating as the ADZ12 but a lower voltage rating.

It was decided to use the same output transformer as for the 28 V circuit, but to redesign the input transformer. The circuit specification and performance are given in Table 8.

It can be seen that the efficiency in this case is much lower. This is mainly due to the lower supply voltage, since the efficiency depends on the ratio of $V_{CE(sat)}$ to V_{cc} and circuit losses are approximately the same.

A 100 *W* 18 *kHz Inverter*

For high-frequency operation, transistors with a higher cut-off frequency f_T are required than those normally available with germanium transistors.

Silicon transistors have an f_T above 10 MHz and therefore are suitable for inverters operating above the audio-frequency range.

A practical example for a 100 W 18 kHz inverter (Refs. 130, 131) using RCA 2N5202 silicon epitaxial n–p–n transistors is shown in Figure 175. Performance characteristics for the inverter are shown in Figure 176.

Summary

The two-transformer circuits described above have distinct advantages over the normal arrangements using a saturable transformer.   Briefly these advantages are (i) improved performance under varying load conditions with reduced stress on the transistors, even though the operation up to the maximum ratings of the transistors is permissible; (ii) transformer design is less critical by allowing the use of conventional transformers rather than the large and expensive saturating transformer.   This results in a 50 or 60% reduction in the cost of the transformers.

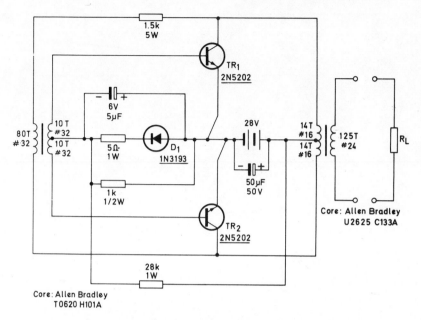

FIGURE 175.   Circuit diagram of 100 W 18 kHz inverter. [By courtesy of R.C.A., Ref. 131.]

Inverter With *CR* Timing

The d.c. inverter described in this section is another circuit from the family of possible arrangements in which base currents are controlled rather than collector currents.   In fact it is an extension of the previously described two-transformer circuit for higher frequencies.   The saturable drive transformer is replaced by a *CR* combination (Refs. 139, 140).   The basic circuit diagram of a square-wave inverter using a single transformer and capacitive–resistive timing is shown in Figure 177.

The timing of the inverter is controlled by the decrease of base current, using the *CR* timing circuit.   It makes the circuit particularly useful for high frequencies ranging from 1 kHz to 100 kHz or higher.

The peak collector current is very nearly equal to the load current, except for a small increase due to the magnetising current of the linear (non-saturable) transformer.

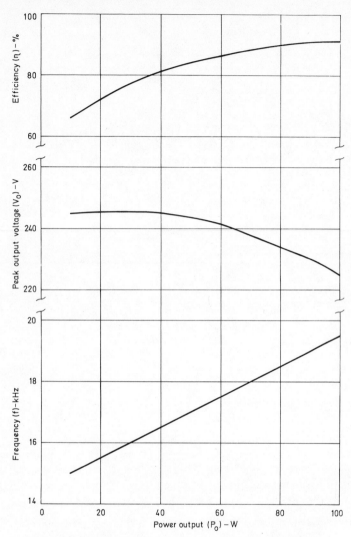

FIGURE 176. Performance characteristics of inverter shown in Figure 175. [By courtesy of R.C.A.· Ref. 113.]

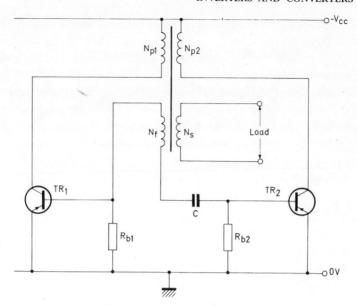

FIGURE 177.   Basic inverter with *CR* timing.

Principles of Operation

In order to describe the operation, it is assumed that the transistor TR_1 is cut off and TR_2 is on.   The capacitor C is charged by the base current of the conducting transistor, its voltage opposing the transformer signal drive.   The base current therefore decreases exponentially until it can no longer support the collector load current.   At this stage the collector current starts to fall, causing the polarities of the voltages induced in all the windings to reverse.   This reversal rapidly switches TR_2 off and TR_1 on.

C, will now charge with the opposite polarity through R_{b2} and the base of TR_1 and the instant of switching is again controlled by the decrease of the base current.   The operating frequency depends on C and on the value

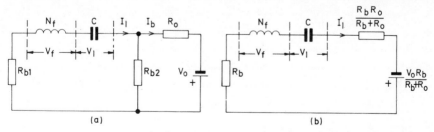

FIGURE 177.   (a) Equivalent base circuit of Figure 177. (b) Simplified equivalent circuit.

of the equivalent-circuit resistor, which is found from the circuit shown in Figure 178(b).

A linear transformer steps up the collector voltage to the value required for operation of the lamp.

The basic circuit of Figure 177 will not start reliably, as no bias is applied to the transistors and the initial gains are low. In a practical inverter circuit, such as Figure 181, two starting resistors are added to supply a little forward bias. When the supply is switched on, the transistors are biased into the active region, and the transistor with the higher gain conducts and starts the operating cycle.

Design Procedure

The most suitable starting point for the design is to consider the base input circuit and timing arrangement. The next step is to determine the primary inductance for a particular magnetising current and then the number of turns required for each winding.

The equivalent base circuit of the basic inverter is given in Figure 178(a), where V_f is the voltage developed across the feedback winding, C is the timing capacitor, and R_{b1}, R_{b2} are the base resistors. Making $R_{b1} = R_{b2} = R_b$, and using Trevenin's theorem, the equivalent circuit may be simplified, as shown in Figure 178(b). The linearised equivalent resistance of the transistor input can be found from the (I_b, V_b) characteristic of the transistor, as shown in Figure 179.

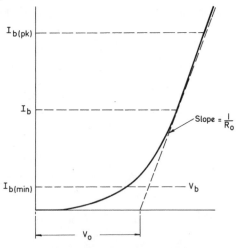

FIGURE 179. Transistor input characteristic.

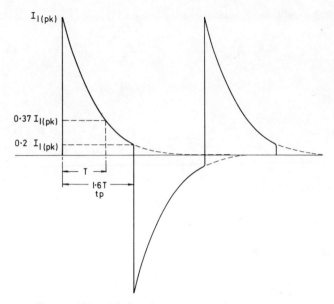

FIGURE 180. Discharging current of timing capacitor.

From Figure 178 the final voltage $V_{1(max)}$ developed across the timing capacitor is found to be

$$V_{1(max)} = V_f - I_{1(min)} \left(R_b + \frac{R_b R_o}{R_b + R_o} \right) - \frac{V_o R_b}{R_b + R_o}, \qquad (181)$$

where

$$I_{1(min)} = I_{b(min)} + \frac{I_{b(min)} R_o + V_o}{R_b}. \qquad (182)$$

The peak current discharging the capacitor is (Figure 180)

$$I_{1(pk)} = \frac{V_f + V_{1(max)} - V_o R_b/(R_b + R_o)}{R_b + R_b R_o/(R_b + R_o)}. \qquad (183)$$

From Equation (183)

$$V_{1(max)} = I_{1(pk)} \left(R_b + \frac{R_b R_o}{R_b + R_o} \right) + \frac{V_o R_b}{R_b + R_c} - V_f. \qquad (184)$$

Equating the right-hand sides of Equations (181) and (184),

$$V_f - I_{1\,(min)}\left(R_b + \frac{R_b R_o}{R_b + R_o}\right) - \frac{V_o R_b}{R_b + R_o} = I_{1\,(pk)}\left(R_b + \frac{R_b R_o}{R_b + R_o}\right) + \frac{V_o R_b}{R_b + R_o} - V_f,$$

from which

$$V_f = \tfrac{1}{2}\left(I_{1\,(pk)} + I_{1\,(min)}\right)\left(R_b + \frac{R_b R_o}{R_b + R_o}\right) + \frac{V_o R_o}{R_b + R_o}. \tag{185}$$

Now, if it is assumed that

$$I_{1\,(pk)} = 5I_{1(min)}, \tag{186}$$

Equation (185) becomes

$$V_f = 3I_{1(min)}\,R_b\left(1 + \frac{R_o}{R_b + R_o}\right) + \frac{V_o R_b}{R_b + R_o}. \tag{187}$$

In this equation, $I_{1\,(min)}$ may be found by substituting in Equation (182):

$$I_{b\,(min)} = \frac{I_{CM}}{h_{FE}}, \tag{188}$$

where I_{CM}, the peak collector current, is approximately equal to the sum of the collector load current $I_{c(load)}$ and the magnetising current I_M of the transformer, that is

$$I_{CM} = I_{c(load)} + I_M. \tag{189}$$

R_o, h_{FE}, and V_o are known for a given transistor type, leaving V_f and R_b to be determined.

The optimum value of R_b, that is the value of R_b for minimum power loss in the input circuit, may be found by multiplying together expressions for voltage and current around the base circuit, and by differentiating the resulting expression for power with respect to R_b.   The timing capacitance C is then chosen to give the correct operating frequency f from

$$f = \frac{1}{2t_p}, \tag{190}$$

where t_p is the time of duration of the pulse, as shown in Figure 180.   The value of t_p may be found from the assumption given in Equation (186), that is from

$$I_{1(pk)} = 5I_{1(min)},$$

since

$$I_1 = I_{1(pk)} \exp\left(\frac{-t_p}{\tau}\right)$$

or

$$\tfrac{1}{5} = \exp\left(\frac{-t_p}{\tau}\right).$$

Therefore

$$t_p = \tau \ln(5)$$
$$= 1 \cdot 6\tau, \tag{191}$$

and the time constant of the circuit, τ, neglecting the resistance of the feedback winding, is given by

$$\tau = C\left(R_b + \frac{R_b R_o}{R_b + R_o}\right)$$

$$= CR_b\left(1 + \frac{R_o}{R_b + R_o}\right). \tag{192}$$

Now by substitution of t_p and τ from Equations (191) and (192) in Equation (190), the operating frequency becomes

$$f = \frac{1}{3 \cdot 2CR_b\left\{1 + R_o/(R_b + R_o)\right\}}. \tag{193}$$

The transformer used for the inverter is a normal linear transformer. If the operating frequency is made high enough and a Ferroxcube core is used, the transformer can be made quite small.   There is only one condition to be satisfied, that the primary windings must have a sufficiently high inductance to keep the magnetising current low.   The inductance required for each half of the primary winding can be found from

$$L_p = (V_{cc} - V_{CE(sat)})\,\frac{t_p}{I_M}, \tag{194}$$

where V_{cc} is the supply voltage $V_{CE(sat)}$ is the collector voltage of the bottomed transistor, and t_p is the time of a half-cycle. The number of primary turns is then determined from the known characteristics of the core material. The number of turns for the feedback winding, N_f, is given by

$$N_f = N_p \frac{V_f}{V_{cc} - V_{CE(sat)}},\qquad(195)$$

where V_f is given by Equation (187).

The number of turns required for the output winding is determined by the output voltage required.

Practical Examples

Inverter Circuit for Two 40 W Fluorescent Lamps

Requirements of the Fluorescent Lamp. The main difficulty in designing an efficient and reliable inverter is in satisfying the requirements of the fluorescent lamp (Ref. 142). These requirements are a high striking voltage, which may have any value between 200 and 500 V, and a low burning voltage of 50 to 150 V, depending on the type of lamp, its power, and length. The lamp behaves like a negative resistance, the voltage across the lamp decreasing for an increase in current, so that the burning voltage is not constant, but depends on the current. A current-limiting device or ballast should therefore be connected in series with the lamp.

Attention should also be given to the voltage applied across the lamp and to the peak value of the current, because of the undesirable consequences that excessive values may have on the life of the lamp. The ratio of the maximum lamp current to the effective value—the peak factor—for 50 Hz sine wave must not exceed 1·7, in order to avoid undue shortening of the lamp life.

Most lamps are now made with filament-heated cathodes and an earthed strip, in order to reduce the ignition voltage. Starting from cold, in general, damages the cathodes and shortens the life of the lamp. In pre-heated lamps, the efficiency is lower because of the power absorbed by the filaments, and is also affected by losses in the ignition strip. This type of lamp was chosen because of the lower striking voltage and longer life. 40-W 4-ft lamps were used, having a starting voltage of about 200 V, a running voltage of 108 V, and heater voltages of approximately 9 V each.

Practical Design of a Two-Lamp Circuit. A practical circuit of the two-lamp inverter is shown in Figure 181 (Refs. 139 to 141).

If only one lamp is used, with inductive ballast, the circuit does not make efficient use of the transistor capabilities because of the approximately triangular waveform of the collector current due to the highly inductive load.

By using two lamps, one with inductive ballast and one with capacitive ballast, the collector current can be made approximately rectangular, thus making the most efficient use of the transistors.

In the circuit diagram given in Figure 181, it can be seen that one heater winding is common to both lamps. The other two heater windings are separate in order to satisfy the different ballast requirements.

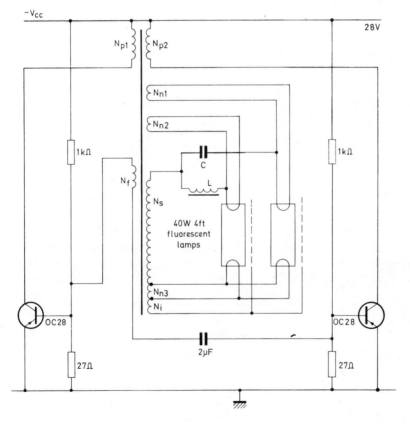

FIGURE 181. Practical inverter circuit. [By courtesy of *Electronic Engineering*, Ref. 140.]

For efficient operation of the inverter the ratio of the magnetising current to the load current should be as small as practically possible. Low magnetising current, however, calls for a high primary inductance which in turn demands large numbers of turns for all windings, for a given size of transformer core. A compromise is therefore necessary, and, in order to realise a design of the circuit, a few assumptions are made.

Assuming a magnetising current I_m for the transformer, of 500 mA, a collector load current $I_{c(load)}$ of 3·5 A, and a typical h_{FE} of the OC28 of 30, then from Equation (189) the peak collector current $I_{c(max)}$ is 4 A, and from Equation (188) the minimum base current $I_{b(min)}$ is 133 mA.

The values of R_o and V_o as estimated from the (I_b, V_b) characteristics of the transistor are 1 Ω and 0·6 V.

The next step is to determine the value of R_b. For practical purposes, R_b should be considered in conjunction with the desired operating frequency and with the timing capacitance C.

Finding R_b by differentiating an expression for the power, as suggested previously, is not suitable here, because either a high timing capacitance or a high operating frequency would be required. A compromise is therefore made. For an operating frequency of about 5 kHz and a timing capacitance of 2 μF, the value of R_b is chosen to be 27 Ω from Equation (193).

$I_{1(min)}$ is then found to be 160 mA using Equation (182), which gives an $I_{1(pk)}$ of 800 mA using Equation (186). The feedback voltage, using Equation (187), is found to be approximately 13·4 V.

The Transformer. For an operating frequency of 5 kHz, t_p is 100 μs, and the primary inductance L, allowing for a drop of 1 V across the transistor, is found to be 5·4 mH, from Equation (194). Using two Ferroxcube E cores type FX1819 the number of turns for each primary winding is given by $N_p = 23(L)^{\frac{1}{2}}$ mH, which gives 53 turns.

The number of feedback turns is then found from Equation (195) to be 26.

It is now necessary to determine the number of turns required for the output windings. The windings include three heater windings N_h, the secondary winding N_s, and an ignition winding N_i, one end of which is connected to the strip.

A transformer was built up from a preliminary design, in order to establish empirically the most suitable values of the voltages for the output windings. From the results of measurements carried out with this transformer, the voltages for each winding were chosen to give satisfactory operation, taking into account the entire spread in transistor characteristics and the variation in supply voltage.

For the heater voltage V_h of 9 V, the secondary voltage V_s of 170 V, and an additional ignition voltage V_i of 80 V, the appropriate numbers of turns are

$$N_h = N_p \frac{V_h}{V_{cc} - V_c} = 53 \frac{9}{28 - 1} \approx 18 \text{ turns},$$

$$N_s = N_p \frac{V_s}{V_{cc} - V_c} = 53 \frac{170}{28 - 1} \approx 334 \text{ turns},$$

and

$$N_i = N_p \frac{V_i}{V_{cc} - V_c} = 53 \frac{80}{28 - 1} \approx 157 \text{ turns}.$$

It should be noted that (as in all push–pull inverters) any leakage inductance between the primaries would cause spikes of voltage at the collectors. The primary windings of the transformer should be bifilar wound to minimize the leakage inductance.

Choke and Capacitor Ballasts. Because of the negative resistance characteristic of the fluorescent lamp, ballast is required for satisfactory operation. The ballast must fulfil the starting requirements for the lamp, and, when the lamp is running, it must pass the correct lamp current at the correct voltage.

To simplify the design of the ballast, only the fundamental component of the square-wave voltage appearing across the secondary of the transformer is considered. The voltage drop across the ballast can be found from a vector diagram. The ballast reactance is then found, taking into account the mean current flowing through the lamp under normal working conditions.

If V_s is the secondary voltage and V_L the voltage across the lamp when it is working, the voltage drop across the ballast reactance, V_x, is given by

$$V_x = (V_s^2 - V_L^2)^{\frac{1}{2}} \tag{196}$$

Substituting $V_s = 170$ V and $V_L = 108$ V, the value of V_x is found to be 131 V. Therefore for a lamp running current of 0·41 A a ballast impedance of 318 Ω is required, and, for an operating frequency of 5 kHz, the ballast inductance should be approximately 10 mH and the ballast capacitance 0·1 μF.

The inductance was obtained with 190 turns of 24 s.w.g. enamelled copper wire wound on Ferroxcube E and I cores (types FX1007 and FX1107) with an air gap of 0·3 mm.

Summary of Transformer and Choke Details

Transformer core material:　　　2 Mullard Ferroxcube E cores FX1819.

Transformer primary winding:　　N_p, 53 turns each of 20 s.w.g. enamelled copper wire (bifilar wound).

Transformer feedback winding: N_f, 26 turns of 26 s.w.g. enamelled copper wire.

Transformer output windings:　N_{h1}, N_{h2}, 18 turns each of 24 s.w.g. enamelled copper wire,

N_{h3}, 18 turns of 22 s.w.g. enamelled copper wire,

N_s, 334 turns of 26 s.w.g. enamelled copper wire,

N_i, 157 turns of 30 s.w.g. enamelled copper wire.

2 Mullard Ferroxcube cores, E and I, FX1007 and FX1107, 190 turns of 24 s.w.g. enamelled copper wire, air gap 0·25 mm.

Circuit Performance.　　The efficiency of the inverter can be measured by comparing the light output of the fluorescent lamp fed from the d.c. inverter with the light output of the lamp energised from the 50 Hz mains.　The efficiency in question is a relative efficiency, and is the ratio of the input power using nominal 220 V 50 Hz mains to the power required for the inverter from a 28 V supply.　For typical transistors, the relative efficiency is about 82% for one 40 W lamp with inductive ballast.　The operating frequency for typical transistors is about 4·3 kHz.　However, the frequency varies with the h_{FE} of the transistors.　The base current for high-h_{FE} transistors will cut off at a much lower value, and take a longer time to discharge the timing capacitor (see Figure 181), so that the frequency will be lower.　For low-h_{FE} transistors, the current will cut off at a higher value, thus making the frequency higher.　Depending on the values of h_{FE} of the transistors, the operating frequency may be between 4 and 5·6 kHz.

The variation of the light output with h_{FE} is not very great, as the collector current depends mostly on the load conditions. The total variation encountered was about 3%.

All the above measurements and observations were made with a supply voltage of 28 V. It is important, however, for the inverter to work reliably at low supply voltages. Figure 182 shows the effect of varying the supply voltage over a wide range on the starting and performance of the fluorescent lamp. The efficiency (light output/power input) improves as the supply voltage is lowered, because the waveform of the current through the lamp changes from an exponential or nearly a sine wave to nearly a square wave.

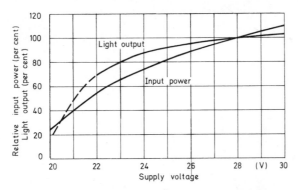

FIGURE 182. The operating characteristic. [By courtesy of *Electronic Engineering*, Ref. 140.]

The above comments apply to the inverter feeding one 40 W lamp with inductive ballast. When two 40 W lamps are used, one with inductive ballast and one with capacitive ballast, the performance of the circuit is improved because of better utilisation of the collector current of the transistors, as mentioned above. The relative efficiency is considerably higher, and is nearly 100%.

By means of suitable switching arrangements one lamp could be used at a time if required. However, because of the large ballast capacitance, excessive peak lamp current may be expected if the lamp with the capacitive ballast is used by itself or with the other lamp disconnected. Although this mode of operation is not catastrophic, it is considered a fault condition, because it could lead to a considerable reduction in the life of the lamp unless the necessary steps are taken to limit the current. If at any time, therefore, only one lamp is required, the lamp with the inductive ballast should be used.

A 25 kHz Inverter for Fluorescent Lamp

The operation of fluorescent lamps at frequencies above the audio range eliminates audible noise, and also a greater light output is obtained at 25 kHz, for example, compared with operation at 50 Hz and the same input power. In addition, a transformer using higher-frequency materials requires a smaller core and fewer turns.   Hence, the transformer losses are smaller, and the volume and the weight of the transformer are reduced.

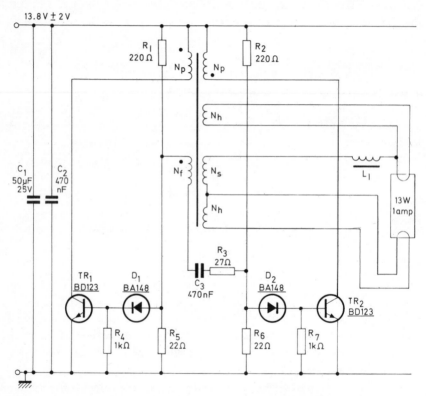

FIGURE 183.   Inverter for flourescent lamps.

An inverter operating from a 13·8 V supply is shown in Figure 183 (Ref. 142).   The output transformer has two heater windings (N_h) for the lamp, in addition to the secondary winding (N_s).   A ballast choke (L_1) is used to limit the current in the lamp.   The frequency of operation is approximately 25 kHz.

Fluorescent lamps suitable for operation from low-voltage supplies are widely available with powers from 4 to 40 W; a 13 W lamp is used in this circuit.

For a case temperature of 60 °C at 25 °C ambient, the transistors each require a heatsink of thermal resistance 4·2 degC/W; this can be obtained by using 200 cm² of 1·625 mm (or 16 s.w.g.) aluminium per transistor.

Transformer Details.

Ferroxcube core:	Mullard FX2242.
Bobbin:	Mullard DT2180.
Board for printed wiring:	Mullard DT2233.
Mounting clip:	Mullard DT2234.
Primary winding − N_p:	12 + 12 turns of 23 s.w.g. enamelled copper wire.
Secondary winding − N_s:	228 turns of 32 s.w.g. enamelled copper wire.
Heater winding − N_h:	9 turns of 30 s.w.g. enamelled copper wire (each).
Feedback winding − N_f:	10 turns of 30 s.w.g. enamelled copper wire.

Choke Details − L_1.

Ferroxcube core:	Mullard FX2240 with 0·09 mm gap.
Bobbin:	Mullard DT2179.
Board for printed wiring:	Mullard DT2227.
Mounting clip:	Mullard DT2228.
Winding:	164 turns of 32 s.w.g. enamelled copper wire.

A 60 W 22 kHz d.c.-to-d.c. Converter

A d.c.-to-d.c. converter operating from a supply of 24 V ± 10% is shown in Figure 184.   The circuit can provide 60 W at direct voltages up to 1000 V if suitable rectifiers are used and the secondary winding is adjusted.   The converter operates at 22·2 kHz, and uses two BD123 transistors.

The upper limit of the supply voltage should not be exceeded if transistor failure is to be avoided; the breakdown voltage of the BD123 is 60 V.   Under cut-off conditions, the collector voltage is twice the supply voltage plus an additional spike.

The supply voltage can be reduced well below 20 V with no loss of performance, except that the output voltage will be proportionally lower.

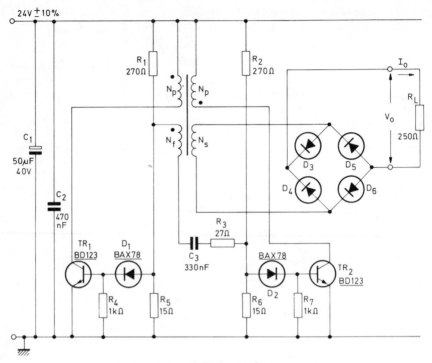

FIGURE 184. 60 W d.c.-to-d.c. converter.

The lower limit is reached when the converter fails to start oscillating under full load conditions. For supply voltages of less than 20 V, the value of R_1 and R_2 should be reduced to 220 Ω or even 180 Ω. Values less than 180 Ω result in a high standing current and consequent loss of efficiency.

The diodes in the bridge circuit should have fast recovery times, otherwise the efficiency will be low, and the transistors can be damaged by the effective short-circuit output during the reverse recovery period. The forward recovery time of the bridge diodes should be less than 90 ns, and the reverse recovery time should be less than 80 ns; they should have a peak inverse voltage of 400 V, and a d.c. forward current rating of 1 A.

Suitable diodes are Dickson 1N4993 to 1N4997 or Unitriode UTX105 to UTX125.

For a case temperature of 60 °C at 25 °C ambient, the transistors each require a heatsink of thermal resistance 4·2 degC/W; this can be obtained by using 200 cm² of 1·625 mm (or 16 s.w.g.) aluminium per transistor.

The measured performance is as follows:

supply voltage	24 V $\pm$ 10%,	load resistance	250 Ω
input current	3·34 A,	output power	60·5 W,
input power	80 W,	efficiency	75·7%
output voltage	121 V,	operating frequency 22·2 kHz,	
output current	500 mA.		

Transformer Details.

Ferroxcube core:	Mullard FX2243.
Bobbin:	Mullard DT2206.
Primary winding $-N_p$:	12 + 12 turns of 21 s.w.g.
Secondary winding $-N_s$:	68 turns of 28 s.w.g. enamelled copper wire.
Feedback winding $-N_f$:	11 turns of 26 s.w.g. enamelled copper wire.

Inverters with *LR* and *LC* Timing

Two other inverters, which belong to the same family (Figures 185 and 186, Refs. 128, 129) as the previously described circuits, are mentioned here. The operation is similar in that a linear (non-saturable) transformer is used for the output circuit and the control is performed at low-power level. The timing arrangement is now replaced either by an *LR* or *LC* combination.

The circuit shown in Figure 185 uses *LR* timing elements. In this case the base current of the conducting transistor will decay exponentially with a time constant approximately equal to $L/2R$. The transistor will switch off when the base current has fallen to I_{CM}/h_{FE}.

Finally, the arrangement shown in Figure 186 uses a *LC* tuned circuit. Because of the starting resistors, which are necessary to bias the transistors into conduction, oscillation will commence and over a number of cycles the energy will build up in the tuned circuit.

A half-wave of base current will flow through the conducting transistor, the *LC* tuned circuit, and the diode connected to the base of the transistor which is cut off (say, TR_1 and D_2). During the other half-cycle TR_1 will be cut off and the base current of the transistor TR_2 will flow through the tuned circuit and diode D_1.

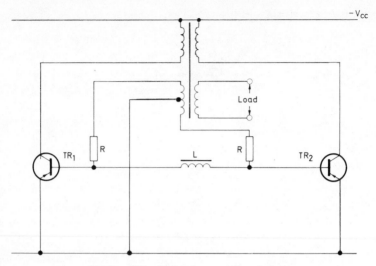

FIGURE 185.   Inductive–resistive timing.

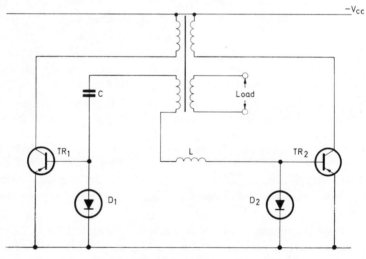

FIGURE 186.   *LC* tuned circuit timing.

The base current is much greater than the value needed for the collector load current and therefore the rise and fall times will be relatively short and the circuit will produce a square-wave output.   The frequency of the oscillation will be approximately constant and will be determined by the *LC* series resonance.

In both circuits described above, the drive is arranged so that the transistors operate either in the bottomed or cut-off condition producing a square-wave output.

Bridge Inverters

When transistor circuits are used with high supply voltages, there is a danger that the maximum safe limits of the transistors may be exceeded. Nevertheless, it is often desirable to operate d.c. inverters from supply rails of higher potential than the transistor ratings will allow.

With a bridge circuit of four transistors (Refs. 138, 144), higher supply voltages can be used.   The peak voltage across any one of the transistors, in the circuits described, does not exceed the supply voltage.

The bridge inverter circuits follow the same pattern as the push–pull arrangements.   In fact, as it is shown in this section, any push–pull circuit can be made into a bridge circuit by adding the necessary two transistors and the feedback windings and by modifying the starting arrangement to allow for the additional transistors and for the higher supply voltage.

As in the case of push–pull circuits, the bridge inverter circuits are divided into two groups: (i) inverters using single saturable transformer where the switching is controlled, at high-power level, in the collector circuit; and (ii) inverters using non-saturable transformers for the output and a separate arrangement for controlling switching at the base power level.

A brief description of various circuit arrangements will follow.   The transformer design and the design of the input timing circuits is the same as for push–pull circuits and therefore will not be repeated here.

Bridge Inverter with Saturable Transformer

The first type of bridge circuit shown in Figure 187 is analogous to the push–pull circuit with saturable transformer.   The circuit is completely symmetrical.   It uses four transistors which conduct in pairs through the primary winding of the transformer (Ref. 144).

Starting Circuit

As with all inverter circuits discussed up to now, the bridge inverter requires a starting circuit.   A suitable starting arrangement is shown in

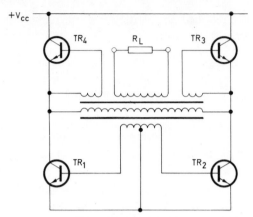

FIGURE 187. Basic bridge inverter circuit
with saturable transformer.

Figure 188. The starting bias is only applied to the bottom two transistors
TR_1 and TR_2 via a centre tap on the primary of the transformer T_1.

Therefore, in order to explain the action of the starting circuit, the
transistors TR_3 and TR_4 can be ignored and the circuit can be redrawn
as shown in Figure 189. This is the effective part of the bridge inverter
at switch-on and, in fact, is the push-pull circuit of Figure 161, with a resistor
R_1 added between the centre tap of the transformer primary and the supply
line V_{cc}.

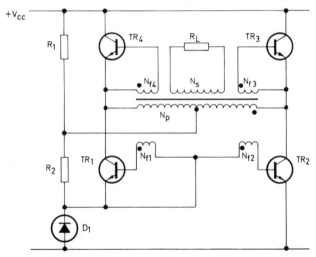

FIGURE 188. Bridge inverter showing starting
arrangement.

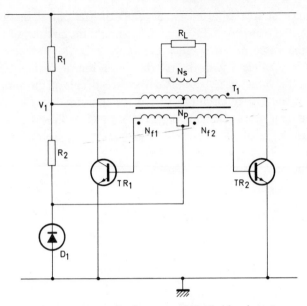

FIGURE 189.   Effective part of the bridge inverter
of Figure 188 at switch-on.

In Figure 189, if a large-value capacitor were connected across the R_2 and D_1 combination and R_1 considered as a dropper resistor, the circuit would behave exactly like the push-pull circuit described previously. However, in the actual bridge inverter circuit, R_1 completes a return path for the *starting* collector currents of the transistors TR_1 and TR_2.

The value of R_2 is therefore given by

$$R_2 = \frac{V_1}{2 I_B + I_{D1}}, \qquad (197)$$

where I_B is the base current of each transistor and I_{D1} is the leakage current of diode D_1.

For silicon diodes I_{D1} can be ignored: therefore

$$R_2 \leqslant \frac{V_1}{2 I_B}. \qquad (198)$$

The value of I_B can be found by considering the loop gain required for oscillation similar to the push-pull inverter.

The choice of the value of V_1 depends on the power dissipation that is allowed in the transistors TR_1 and TR_2 should the circuit fail to oscillate. Therefore the value of V_1 should preferably be as low as possible whilst still providing sufficient gain for reliable initiation of oscillation. On the other hand, the value of V_1 should never exceed half of the supply voltage V_{cc} used for the bridge circuit. This is the limiting case for converting a push-pull circuit to a bridge configuration when $V_1 = V_{cc}/2$, otherwise, the values of V_1 used in practical bridge circuits are found to be between $V_{cc}/4$ and $V_{cc}/3$.

The value of R_1 is simply given by

$$R_1 = \frac{V_{cc} - V_1}{2I_E}, \tag{199}$$

where I_E is the emitter current of each transistor.

The resistors R_1 and R_2 are used for the following two reasons: (i) to provide forward bias to transistors TR_1 and TR_2 in order to start the oscillation, and (ii) to provide a path for the collector currents of the transistors at the moment of starting.

Circuit Operation

Assume that, owing to imbalance in the circuit, the transistor TR_1 conducts more heavily than TR_2 on the initial switch-on, and sets the polarities of the induced voltages as indicated in Figure 188. The voltage developed across the base winding N_{f1} makes the transistor TR_1 go further into conduction, at the same time keeping the transistor TR_2 cut off. This is a cumulative action leading to the bottoming condition of the transistor TR_1. The voltages set up in the remaining windings of the transformer, due to the polarities indicated by the dots in Figure 188, force the transistor TR_3 to conduct and the transistor TR_4 to cut off. Thus the magnetising current, initially flowing in one-half of the primary winding, continues to flow in the other half of the winding and through the transistor TR_3, provided that sufficient voltage is developed across its base winding N_{f3}.

The transistors TR_1 and TR_3 continue to conduct until the transformer saturates. At this stage the magnetising field collapses causing reversal of polarity in all the windings and thus switching the transistors TR_2 and TR_4 on, and the transistors TR_1 and TR_3 off. The magnetising current now flows in the opposite direction through the primary winding until negative saturation is reached. This completes the cycle and the action is repeated as long as the supply is connected to the circuit. The conduction period depends on the amount of feedback voltage applied and on the saturation characteristics of the transformer core. The peak current through the primary winding is the lesser of the peak currents available from the conducting transistors. This can be seen easily by considering the individual transistors. As one transistor fails to supply the increasing magnetising current required to maintain the voltage across the transformer, it comes out of saturation and resulting regeneration causes the circuit to switch over.

The current flows through diagonally opposite transistors TR_1 and TR_3 and the primary winding in one direction, and then reverses and flows through TR_2 and TR_4 and the primary winding in the other direction. Therefore, when TR_1 and TR_3 are conducting, the supply voltage less $V_{CE(sat)1} + V_{CE(sat)2}$ appears across the primary winding. The supply voltage also appears across the other two transistors TR_2 and TR_4, which are cut off, so that the voltage across any transistor will never exceed the supply voltage V_{cc}. Hence this inverter can be used with a supply voltage of twice the value allowed for any push–pull arrangement using similar devices.

On the assumption of equal bottoming voltages, the square-wave amplitude of the voltage across the primary is

$$V_p = V_{cc} - 2V_{CE(sat)} - V_{Rp},\tag{200}$$

where V_{cc} is the supply voltage, $V_{CE(sat)}$ the collector-to-emitter saturation voltage, and V_{Rp} is the voltage drop across the resistance of the primary winding.

The transformer can now be designed using a similar approach to that used in the case of the push–pull inverter (page 198).

The four feedback windings are arranged to give sufficient drive such that, under the worst conditions, the collector current equals the maximum load current referred to the transformer primary winding. The drive to transistors TR_1 and TR_2 should be 0·5 V higher than the value for the worst conditions, to compensate for the drop across the starting diode.

It is usual practice to add a small resistance in series with each base to reduce the effects of spreads in the transistor input characteristics.

Bridge Inverter with Two Transformers

The bridge inverter with two transformers (Ref. 128) operates on the same principle as the push-pull two-transformer circuit described previously. The advantages of both bridge and two-transformer circuits are thus combined.

The basic circuit is shown in Figure 190 and a practical circuit with a suitable starting arrangement is shown in Figure 191.

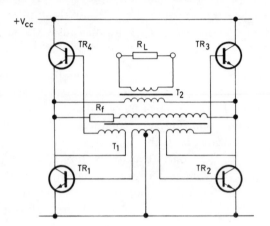

FIGURE 190. Basic bridge inverter circuit with two transformers.

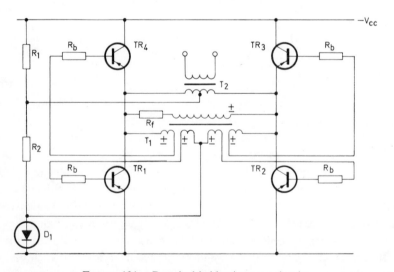

FIGURE 191. Practical bridge inverter circuit.

The required drive is applied by a small saturating transformer T_1, in conjunction with a feedback resistor R_f. The polarities of the transformer windings are marked in the diagrams. Transformer T_2 has a linear characteristic and it is used to step up the voltage to the value required for the load.

Diagonally opposite transistors (TR_1 and TR_3, or TR_2 and TR_4) conduct together so that, when TR_1 and TR_3 are bottomed, the supply voltage will appear across TR_2 and TR_4 which are cut off. Therefore the voltage across any transistor will never exceed the supply voltage V_{cc}, and this converter can be used with a supply voltage of twice the value allowed for any push-pull arrangement using similar transistors.

Another advantage is that the converter is suitable for variable loads. This is because the collector current in any transistor does not rise to the peak value determined by the drive but, as in the push-pull circuit described previously (page 205), to a value equal to the sum of the load current, the magnetising current of the output transformer, and the feedbeck current.

This converter has good regulation. It is more economical in construction than the bridge circuit described in the preceding section because it uses only a small saturable transformer which requires little expensive core material. The output transformer, being a conventional type, is relatively cheap for the output power obtained.

OC28 transistors can be used in the suggested circuit, with a 56 V supply, and it is possible to obtain output powers up to 200 W with an overall efficiency greater than 80%. The same transformer design procedure may be adopted as for the earlier circuits. However, the circuit designer must still ensure that the published maximum ratings for the transistors are never exceeded.

Bridge Inverter With CR Timing

A basic bridge inverter circuit with CR timing is shown in Figure 192 (Ref. 145). Only one non-saturable transformer is used. The magnetising current of the transformer is low compared with the load current, and the frequency of oscillation is controlled by means of the CR network connecting the base circuits of the transistors TR_1 and TR_2. N_f is the main feedback winding, whereas N_3 and N_4 are the supplementary feedback windings required to provide the necessary drive for the transistors TR_3 and TR_4 respectively. The design of the feedback windings, in this case, is different from the bridge circuit with two transformers. The main difference is that only three windings are necessary. The winding N_f,

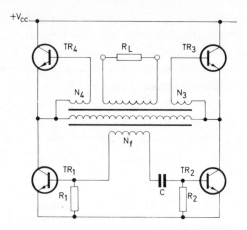

FIGURE 192.   Basic bridge inverter with
CR timing.

which controls the timing of the inverter, is designed on similar lines to
the push-pull inverter with CR timing, whereas the supplementary windings
N_3 and N_4 follow the same lines as the previously described bridge circuits.

A bridge inverter circuit with CR timing complete with biasing arrangement to facilitate starting is shown in Figure 193.

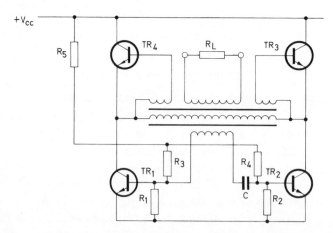

FIGURE 193.   Bridge inverter with CR timing complete
with biasing arrangement.

The values of the starting components are determined by splitting the bias chain into two sections: (i) resistor starting for the transistors TR_1 and TR_2 with V_1 as the d.c. supply voltage, and (ii) the voltage drop across R_5, so that

$$R_5 = \frac{V_{cc} - V_1}{I_5},$$

where I_5 is the current flowing through R_5.

The inverter circuit is particularly useful for high-frequency operation from 1 kHz to 100 kHz with possible extension to the megahertz region.

The circuit offers further advantage over the two-bridge circuits described above; the transformer is simpler as it requires one less feedback winding. Thus complexity is reduced and a price advantage over other configurations may be offered.

Bridge Inverters With *LR* and *LC* Timing

At least two bridge inverter circuits are worth mentioning (Ref. 145). These are a bridge inverter with *LR* timing shown in Figure 194 and a bridge inverter with *LC* tuned circuit timing shown in Figure 195. Both circuits shown are complete with starting bias arrangements. The circuits are practical counterparts of the equivalent push–pull arrangements and, although practical, they have doubtful advantages over the circuits already described. They are included for completeness.

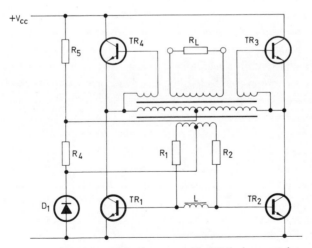

FIGURE 194.   Bridge inverter with *LR* timing complete with starting bias.

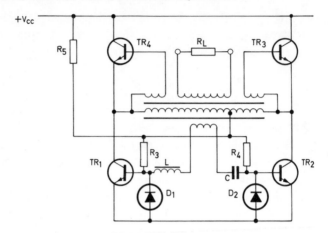

FIGURE 195.   Bridge inverter with *LC* tuned circuit
timing complete with starting bias.

Other Inverter Circuits

Push-Pull Inverter with Current Limiting

Push-pull circuits in their basic form do not include overload protection
facilities.   When overload occurs owing to a short-circuit, very often the
circuits fail to oscillate because the loop gain is reduced below one.   If
an excessive drive is applied or the load current is increased gradually,
the collector current will increase until transistor breakdown occurs.

To include a protection circuit, a certain penalty has to be paid, either
by increasing overall cost or by reducing the efficiency.   For the inverter
efficiency to remain unaffected, the overload protection circuit has to operate
at low-power level; therefore a more complex circuit is required, demanding
more transistors and other components.

If reduction in efficiency is not objectionable, an overload protection
facility can be achieved by the addition of only one extra resistor (Ref. 146).

A basic circuit with such protection is shown in Figure 196.   The resistor
R_E connected between the junction of the two emitters and earth not only
allows close control of the peak collector current of the transistors but also
provides one of the most effective means of balancing the collector currents
due to the spreads in the transistor gain and V_{BE}.   In addition the resistor
provides negative feedback to the oscillatory currents and therefore stabilises
the frequency of oscillation.   This is particularly important when very-

low-frequency operation is required, and also if a wide variation in load current is expected.

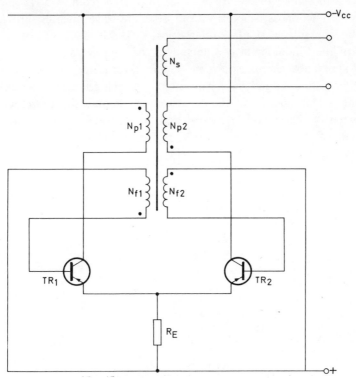

FIGURE 196.   Push-pull inverter with current limiting.

Moreover, the resistor R_E reduces effective spikes of voltage and current which might otherwise appear owing to leakage inductance or highly inductive loads.

The inverter is particularly useful for operation from a car battery and as a practical example, an electric shaver is used for a load.

Design Considerations

The design of the inverter is based on the available supply voltage, which in this case is a car battery.   This, however, may vary from 10 to 15 V so that a nominal value of 12·6 V is taken as a design centre.

The circuit is to operate at 50 Hz producing a square-wave output voltage equivalent to the 230 V a.c. mains. Although electric shavers for use with mains voltage are designed to operate from a 50 Hz sine wave, they will function satisfactorily from a square-wave inverter, provided an adjustment is made to allow for the loss of power due to the lower peak value of the fundamental component of the square wave.

In a practical circuit shown in Figure 197, in addition to the normal starting resistors R_1 and R_2, a capacitor C_1 is included across the secondary winding. The function of the capacitor is to reduce the collector voltage overswing of the cut-off transistor and, at the same time, to slow down the edges of the voltage waveform by bypassing higher harmonics.

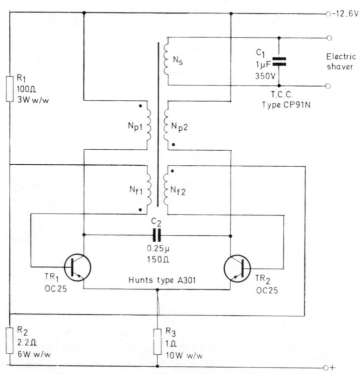

FIGURE 197. Practical circuit.

Choice of Transistor Type

Collector Circuit. To allow a wide range of shavers to be used an upper limit of power capability of the inverter is set at 15 W.

The peak collector current I_{CM} can be estimated from the following expressions:

$$I_{CM} = I_R + I_M + I_f, \qquad (201)$$

where I_R is the collector load current, I_M is the magnetising current of the transformer, and I_f is the feedback current required for oscillation which is small enough to be neglected. Therefore

$$I_{CM} \approx I_R + I_M. \qquad (202)$$

From the knowledge of the supply voltage V_{cc}, the output power P_{out}, and the efficiency η the value of the collector load current can be found from

$$I_R = \frac{P_{out}}{\eta(V_{cc} - V_{CE(sat)})}, \qquad (203)$$

where $V_{CE(sat)}$ is the bottoming voltage of the transistors, say, 0·6 V.

Owing to the fact that an emitter resistor R_3 is employed, the overall efficiency will not be high. If the efficiency is assumed to be about 50%, the collector load current is

$$I_R = \frac{15}{0·5(12·6 - 0·6)} = 2·5 \text{ A}.$$

Another factor affecting the efficiency is the magnetising current. For efficient operation the ratio of the magnetising current of the transformer to the collector load current should be small. This, however, calls for high primary inductance and results in a large transformer. Therefore a compromise is made by letting the magnetising current be a fifth of the load current, which in this case is 0·5 A.

Therefore the peak collector current I_{CM} becomes 3 A.

Collector Voltage. Another consideration is the collector voltage rating of the transistors under cut-off conditions.

The peak collector voltage of either transistor, when cut off, is approximately twice the supply voltage. If the maximum value of the supply voltage is 15 V the peak collector voltage becomes 30 V.

As in all applications, it is good practice to allow a safety margin, and therefore the minimum transistor ratings sought should be I_{CM} of 4 A and V_{CEM} of 40 V.

A recommended economical transistor for this application is the OC25 which satisfies the above requirements.

Transformer Design

Before the number of primary turns can be calculated, it is necessary to fix the value of the emitter resistor R_3 which will determine the available voltage across the primary, V_p. This in turn will determine the number of turns for the primary winding, N_p, as given by

$$N_p = \frac{V_p \times 10^8}{4fAB_{(sat)}}, \tag{204}$$

where f is the frequency of oscillation in hertz, A is the cross-sectional area in square centimetres, and $B_{(sat)}$ is the flux density in gauss.

The shared emitter resistor R_3 performs at least two functions. It limits the peak collector current to the value chosen. It also compensates for the variation between the gains of individual transistors. The higher the value of the resistor used, the better is the balancing effect on peak collector current of the two transistors. However, with a high value of resistor, there is high voltage drop across it, so that a lower voltage is available across the primary resulting in high turns ratios. Therefore a compromise is necessary in choosing the value of R_3. From practical experience 25% of the available supply voltage is a good value to choose for the voltage drop across R_3.

The available voltage across the primary is given by

$$V_p = V_{cc} - V_{CE(sat)} - V_{R_3}. \tag{205}$$

The number of turns for the feedback winding can be found from

$$N_f = N_p \frac{V_f}{V_p}, \tag{206}$$

where

$$V_f = V_{BE} + V_{R_3} - V_{R_2} + I_B \times R_2 \tag{207}$$

and V_{BE} is the maximum base-emitter voltage necessary to drive all transistors to the required collector current and V_{R_2} is the d.c. starting bias.

Since the maximum V_{BE} value is used in order to cover full transistor spreads, the voltage drop across R_2 due to the base current is insignificant in comparison and may be neglected.

Therefore V_f becomes

$$V_f = V_{BE} + V_{R_3} - V_{R_2}. \tag{208}$$

The number of turns needed for the secondary winding is determined by the output voltage required.

$$N_s = N_p \frac{V_s}{V_p}. \tag{209}$$

Practical Circuit

The transformer is designed to saturate at 0·5 A with no load applied, as assumed previously. On full load, which in the case of an electric shaver is mostly inductive, the collector current is allowed to reach the maximum value of 3 A before saturation is reached.

V_p is found from Equation (205) allowing 3 V for V_{R_1}. The value of R_3 is 1 Ω and

$$V_p = 12\cdot 6 - 0\cdot 6 - 3 = 9 \text{ V}.$$

Therefore from Equation (204)

$$N_p = \frac{9 \times 10^8}{4 \times 50 \times 4\cdot 03 \times 15 \times 10^3} = 75 \text{ turns}$$

using a silicon iron core pattern No.12 (I.S.C.O. 403A) of 2·54 cm stack; $A = 4\cdot 03$ cm^2, and $B_{(sat)} = 15\,000$ G.

The feedback voltage V_f from Equation (207) is

$$V_f = 1\cdot 5 + 3 - 0\cdot 3 = 4\cdot 2 \text{ V},$$

so that

$$N_f = 75 \frac{4 \cdot 2}{9} = 35 \text{ turns.}$$

The inverter provides a square-wave output; therefore a correction factor should be applied for the output voltage to be equivalent to the 230 V a.c. mains, otherwise a slight loss of power may be expected.

From Fourier analysis of a square wave the peak value of the fundamental is $4/\pi$, and, since the r.m.s. value of a sine wave is $1/(2)^{\frac{1}{2}}$ of the peak value, the square-wave voltage must equal the rated sine-wave voltage times the factor $\pi/4 \times (2)^{\frac{1}{2}} = 1 \cdot 11$ to have the same power effect. Hence, for the output voltage to be equivalent to the 230 V r.m.s. sine wave, a square wave of 255 V is required.

Therefore the number of turns required for the secondary winding is

$$N_s = 75 \frac{255}{9} = 2130 \text{ turns.}$$

It was found necessary to connect a capacitor C_1 across the secondary winding to protect the transistors against excessive voltage spikes. The capacitor has the effect of slowing down the edges of the square wave and also helps to stabilise the frequency of operation, so that there is very little variation between open-circuited load and full load conditions. The value of the capacitor used was 1 μF.

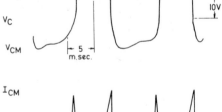

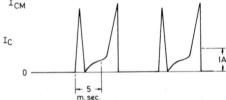

FIGURE 198. Waveforms of collector voltage and current for open-circuit output conditions. [By courtesy of *Wireless World·* Ref. 146.]

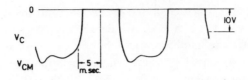

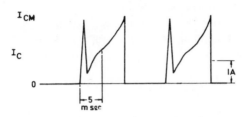

FIGURE 199. Waveforms of collector
voltage and current with shaver connected
across the output. [By courtesy of
Wireless World· Ref. 146.]

The waveforms of the collector voltage and current for no load and
with an electric shaver are given in Figure 198 and 199 respectively.

Using a 12·6 V supply the current drawn without load was 1·12 A.
A number of representative shavers were run from the circuit and the current
increased to between 1·8 and 1·9 A.

To complete the design, the power dissipation in the transistors should
be considered.

The power dissipated in each transistor is estimated on the assumption
of a rectangular current waveform of 3 A and a 1 V drop across the transistor
when it is on and bottomed. This is far in excess of the actual power dissipated which, under the conditions stated, is 3 W over a half-cycle, giving
an average power of 1·5 W.

A typical value of thermal resistance for the transistor used in free air
is approximately 30 degC/W and the maximum junction temperature is
is 90 °C. This allows for operation up to an ambient temperature of 45 °C.

The temperature stabilisation time, however, depends on the thermal
capacity of the device, and the box in which the inverter is enclosed. If
the transistors are mounted in the box, normal free-air figures for thermal
resistance do not apply, and the presence of other heat sources may increase
the ambient temperature inside the box considerably above the assumed
value of 45 °C.

Therefore the transistors should be mounted on the outside of the box and insulated with mica or PVC washers if the box is made of metal, in which case sufficient cooling would be provided. Using a plastic box with transistors mounted on the outside, operation up to 15 min can be allowed at an ambient temperature of 45 °C. This may be marginal in some cases, so that some form of heatsink is recommended for safety. A piece of aluminium, 30 cm^2 × 1·625 mm, with a thermal resistance of approximately 10 degC/W could be used for each transistor. Since the junction temperature rise above the mounting base of the transistor is 2 degC/W, the maximum ambient temperature which could be allowed is

$$T_{amb} = T_j - P(R_{th(j-mb)} + R_{th(h)})$$
$$= 90 - 1·5(2 + 10) = 72 \text{ °C},$$

which is more than adequate and therefore ensures sufficient safety margin.

Transformer Details

Core:	2·54 cm. stack of Linton and Hirst, Pattern No. 12A, 0·020 in Lamcor 3, silicon iron laminations; Joseph Sankey, Pattern No. 102, 42 quality; or M.E.A. Pattern No. 12A (I.S.C.O. 403) Silicor 17, or equivalent.
Bobbin:	Supplied by Armand Taylor Ltd., Pitsea, Essex.
Primary winding:	$N_{p1} = N_{p2}$, 75 turns each of 22 s.w.g. Lewmex copper wire (bifilar wound).
Secondary winding:	N_s, 2130 turns of 36 s.w.g. Lewmex copper wire.
Feedback winding:	$N_{f1} = N_{f2}$, 35 turns each of 34 s.w.g. Lewmex copper wire.

A High-voltage Dual-Transformer Inverter

A dual-transformer inverter (Ref. 133) designed to operate from a 120 V d.c. supply and capable of over 400 W output is illustrated in Figure 200. The input voltage is divided equally across the four series primaries subjecting each transistor to only 60 V in the off condition.

The output voltage is a 60 Hz 140 V square wave. The operating characteristics are illustrated in Figure 201.

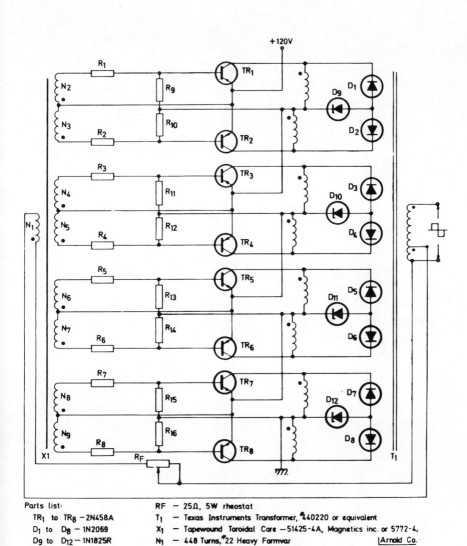

Parts list:

TR₁ to TR₈ — 2N458A	RF — 25Ω, 5W rheostat
D₁ to D₈ — 1N2069	T₁ — Texas Instruments Transformer, #440220 or equivalent
D₉ to D₁₂ — 1N1825R	X₁ — Tapewound Toroidal Core —51425-4A, Magnetics inc. or 5772-4,
R₁ to R₈ — 5Ω, 1W	N₁ — 448 Turns, #22 Heavy Formvar Arnold Co.
R₉ to R₁₆ — 910Ω, 1W	N₂ to N₉ — 112 turns, #28 Heavy Formvar

FIGURE 200.   400-W 60 cycle dual transformer inverter. [By courtesy of Texas Instruments, Inc., Ref. 133.]

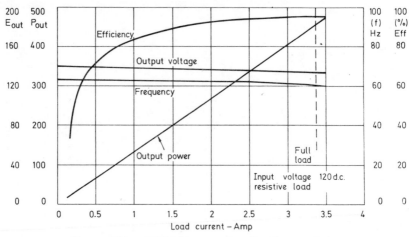

FIGURE 201.   400-W inverter operating characteristics.
[By courtesy of Texas Instruments Inc., Ref. 133.]

Tunnel Diode Converters

Inverters and converters using transistors require power sources which are greater than 1 V. However, a number of power sources exist which have output voltages considerably lower. These are thermoelectric, thermionic and chemical fuel cells, silicon solar cells, and others.

Now that high-current tunnel diodes are available, it is possible to construct what are possibly the simplest d.c. converter circuits yet, operating from the above-mentioned power sources which are only a fraction of one volt.

Brief descriptions of two basic tunnel diode circuits follow.

Single Tunnel Diode Converter

A single tunnel diode converter (Ref. 52) is shown in Figure 202. This is the simplest converter circuit described in this book. The circuit is complete as shown. It consists of a d.c. supply voltage V_{in}, the tunnel diode D_1, and a transformer T_1. No starting or other auxiliary components are required. The transformer T_1 has a single primary N_p and a centre-tapped secondary winding. Two ordinary diodes D_2 and D_3 form a full-wave rectifier circuit feeding an output capacitor C, which provides smoothing, and the load resistor R_L.

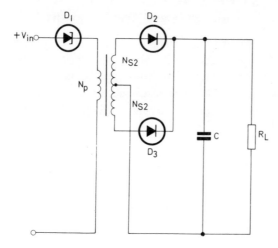

FIGURE 202.   Single tunnel diode converter.

With the supply voltage V_{in} connected as shown in Figure 202, the current that flows in the tunnel diode will start the oscillation due to the negative resistance characteristic of the diode.   The required condition for oscillation is that the d.c. load line intersects the negative resistance region of the diode characteristics at a single point somewhere between the peak and the valley points.

Reference is made to the tunnel diode characteristic shown in Figure 24. Starting from zero, the diode current increases until the peak point is reached. Further increase in the diode current is not possible.   The induced voltage in the transformer winding reverses, biasing the diode in the reverse direction. The operating point shifts to the right of the valley point of the characteristic, and the point on the curve which is level with the peak point is reached. The diode current then decays to the valley point.   The reverse voltage disappears, and the operating point of the tunnel diode shifts back to the origin.   The applied voltage reappears across the diode which causes the current to flow in the forward direction, thus repeating the oscillatory cycle.

With the secondary winding open-circuited, the tunnel diode current is equal to the magnetising current of the transformer.   When the load is connected across the secondary, the diode current is the sum of the reflected load current and the magnetising current.   Either a saturable or a linear transformer can be used.

Push-pull Tunnel Diode Converter

The circuit shown in Figure 203 consists of two tunnel diodes D_1 and D_2 connected in push-pull across the centre-tapped primary of the transformer T_1. The secondary winding is also centre-tapped and is connected to diodes D_3 and D_4, forming a full-wave rectifying circuit. Alternatively, a single winding and a bridge rectifying circuit could be used.

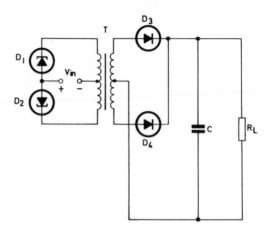

FIGURE 203   Push-pull tunnel diode converter.

The push-pull circuit depends for its operation on the tunnel diode characteristic in a similar way to the single-diode circuit described previously.

The circuit will operate either with a saturable transformer or with a linear non-saturable transformer. The choice of the transformer type depends on frequency stability, efficiency, core weight and size, and the power levels as well as overall cost.

More information covering tunnel diode converters can be found in Refs. 52, 53, 147 to 150.

Three-phase d.c.-a.c. Inverters

The use of transistors in single-ended, push-pull, and bridge inverter circuits described so far in this chapter refer to single-phase operation. For some applications, however, three-phase supplies are required.

A number of methods for generation of three-phase voltages have been described in the literature (Refs. 151 to 155), from which three basic arrangements will be described here.

Three Single-phase Inverters Locked To Provide Three-phase Output

In general any three single-phase inverters of the same type, described previously, can be connected to provide a three-phase output voltage. Different methods are used to achieve the required 120° separation between phases.   These include the use of supplementary reactors for synchronising purposes or special transformer tapping arrangements.   As an example of such, the latter system is shown in Figure 204.

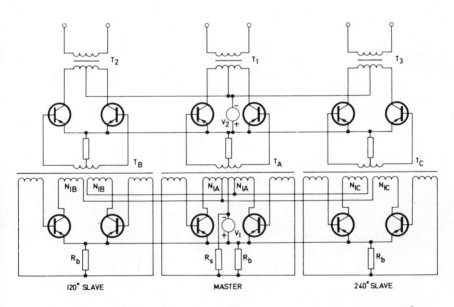

FIGURE 204.   Self-locking three-phase inverter. [By courtesy of A.I.E.E., Ref. 153.]

The circuit consists of three square-wave inverters, a master self-oscillating inverter, and two slaves driving three switched output stages. By virtue of the transformer tapping arrangement the waveforms of the slaves are locked at 120° and 240°, that is, 120° lead and 120° lag with respect to the master oscillator waveform.

Driven, Three- phase *RC* Oscillator, Inverter

An inverter system which is frequently used for generation of three-phase waveforms is shown in Figure 205 (Ref. 154). The system consists of a three-phase *RC* oscillator coupled so that 120° phase difference exists between the transistors TR_1, TR_2, and TR_3. The output from each phase is fed via an emitter follower amplifier to drive the output power transistors which are operated in the saturated switched mode.

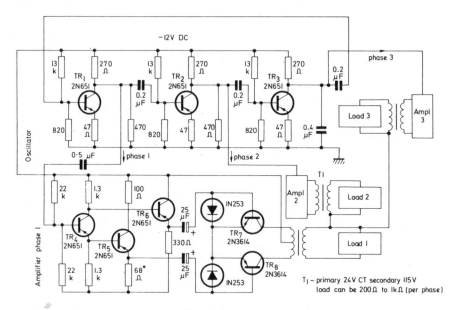

FIGURE 205. 20-W three-phase inverter, 12 V d.c. to 115 V 400 Hz a.c. [By courtesy of Motorola Inc., Ref. 154.]

Only the phase 1 amplifier and its output stage are shown in Figure 205. Identical circuits are used for amplifying phase 2 and phase 3 waveforms.

Balanced Three-phase Self-oscillating Inverter

Another required source of a.c. power is the three-phase supply which is balanced, that is, a supply which has the instantaneous sum of its output voltages always zero. The design of such a supply presents some problems in the use of transistors in the switching mode, since the ideal system of three 1 : 1 square waves displaced by 120° in phase is not balanced (Refs. 151 to 154).

However, a 2 : 1 square-wave system (Figure 206) is balanced, and the generation of a c. power having this waveform will now be discussed (Ref. 155).

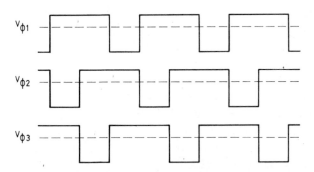

FIGURE 206.   Voltage waveforms of the three output
phases of a 2 : 1 square-wave system.

The oscillator system described uses three transistors operating in a switching mode in which two transistors are on and one is off. Such a system has some advantage over the push-pull system in d.c.-d.c. conversion in that each transistor can handle approximately 30% more power and is therefore used more economically. There are, however, other considerations, particularly that of peak voltage, which limit the usefulness of the system in this respect.

Principle of Operation

Switching Operation. The operation of the three-phase oscillator is basically similar to that of the well known push-pull types; that is the transistor is held bottomed by constant drive to the base while the collector current rises because of the rise in magnetising current in the collector winding. When the collector current can no longer maintain the rising current which is required to keep the voltage across the load constant, this voltage falls, and regenerative action causes a rapid switch-off. The rapid change of flux produced in the core causes the transistor that was previously switched off to come into conduction, and the cycle recommences.

Frequency of Oscillation. As in the push-pull case, frequency stability is achieved by using a transformer with a saturating core. The frequency of operation is determined by the time taken for one limb of the core to change from one saturated state to the other. This time is given by the formula

$$T = \frac{2NAB_s}{V \cdot \times 10^8} . \tag{210}$$

The derivation of this formula will be found on page 234. Figure 207 shows the state of the core after each third of the cycle in the three-phase system described. It can be seen that one-half of each limb moves from one saturated state to the other in one-third of a cycle of oscillation. The frequency of operation is therefore

$$f = \frac{E10^8}{3NAB_s}, \tag{211}$$

where E is the supply voltage, N is the number of turns, A is the cross-sectional area of each limb in square centimetres, and B_s is the saturation flux density of the core material in gauss.

Practical Considerations

Peak Voltage. From the nature of the waveform appearing at the collector, it will be seen that even with perfect flux coupling between the windings the peak voltage appearing at the collectors will be three times the supply voltage. However, the coupling between windings is far from perfect, and very high transient voltages appear at the collectors following switch-off. These can be clipped by diode circuits, as shown in the circuit diagram, Figure 208.

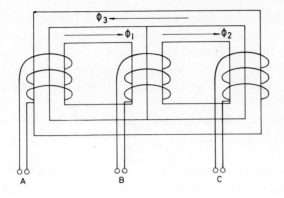

ϕ_1	ϕ_s	$-\phi_s$	ϕ_s
ϕ_2	ϕ_s	ϕ_s	$-\phi_s$
ϕ_3	$-\phi_s$	ϕ_s	ϕ_s

FIGURE 207.   States of flux in the core just before switching, at each third of the cycle.

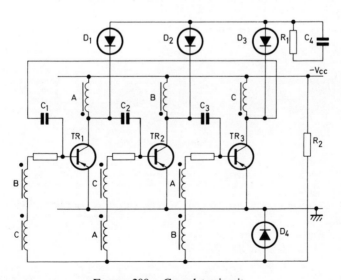

FIGURE 208.   Complete circuit.

Feedback Windings.   If the feedback to each transistor is taken from a winding coupled to its own collector circuit, the switching mode is controlled purely by flux coupling.   Since the coupling is poor, switching tends to be erratic, and the correct sequence is not maintained.   This can be overcome by obtaining the feedback from windings on the other two limbs of the transformer, as shown in Figure 208.   Since the sum of the voltages across each phase winding is zero, the correct feedback is obtained, but only if the switching mode is correct.

Direction of Rotation.   In d.c.-d.c. conversion, the direction of rotation of the switching sequence is immaterial, but, where an a.c. output is required, a knowledge of the phase relationships may be essential.   The switching sequence may be given a preferred direction of rotation by connecting suitable capacitors between the collector of each transistor and the base of the next, as shown in Figure 208.

Starting Circuit.   The diode starting circuit shown is identical in form and operation with that commonly used for push-pull inverters.

Some Properties of the Output Waveform

The form of the phase and between-phase voltages is shown in Figure 209.

Phase Voltages.   The r.m.s. value V of this waveform is given by

$$(\bar{V}\phi)^2 = \frac{1}{T} \int_0^T (V\phi)^2 \, dt = \frac{1}{T}\left(\frac{2E^2T}{3} + \frac{4E^2T}{3}\right) = 2E^2$$

$$\bar{V}\phi = E(2)^{\frac{1}{2}} \tag{212}$$

where $E = -V_{cc}$ in Figure 208.

The harmonic content of the waveform may be obtained from the expression

$$V = \frac{6E}{\pi} \sum_{n=1}^{\omega} \frac{1}{n} \sin\left(\frac{n\pi}{3}\right) \cos(n\omega t). \tag{213}$$

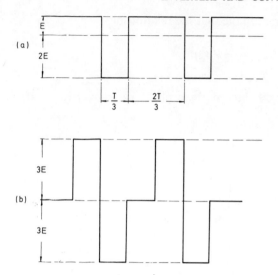

FIGURE 209.   (a) phase voltage; (b) between-phase voltage.

The peak amplitudes of the harmonics are therefore

fundamental $\dfrac{3(3)^{\frac{1}{2}}E}{\pi}$,   fourth $\dfrac{3(3)^{\frac{1}{2}}E}{4\pi}$,

second $\dfrac{3(3)^{\frac{1}{2}}E}{2\pi}$,   fifth $\dfrac{3(3)^{\frac{1}{2}}E}{5\pi}$,

third 0 ,   sixth $0.$

The power in the harmonics as a fraction of the total power available (if we assume a resistive load) is

fundamental 68·4%,   fourth   4·3 %
second   17·1%,   fifth and higher 10·2%.

Between-phase Voltages (Delta Connection).   The r.m.s. value V_δ of this voltage is given by

$$\overline{V}_\delta^2 = \frac{1}{T}\int_0^T V_\delta^2 \, dt = \frac{1}{T}\left(\frac{18E^2 T}{3}\right) = 6E^2$$

$$\overline{V}_\delta = E(6)^{\frac{1}{2}}. \tag{214}$$

It is interesting to note that the relationship

$$\bar{V}_\delta = V_\phi (3)^{\frac{1}{2}}. \tag{215}$$

holds for any balanced three-phase system whatever the waveform. By definition

$$V_\delta = V_{\phi_1} - V_{\phi_2} \tag{216}$$

and

$$\bar{V}_\delta{}^2 = \frac{1}{T}\int_0^T (V_{\phi_1} - V_{\phi_2})^2 dt.$$

$$= \frac{1}{T}\int_0^T V_{\phi_1}{}^2 dt + \frac{1}{T}\int_0^T V_{\phi_2}{}^2 dt - \frac{2}{T}\int V_{\phi_1} V_{\phi_2} dt$$

$$= 2\bar{V}_\phi{}^2 - \frac{2}{T}\int_0^T V_{\phi_1} V_{\phi_2} dt. \tag{217}$$

Since the system is balanced,

$$V_{\phi_1} + V_{\phi_2} + V_{\phi_3} = 0; \tag{218}$$

therefore

$$V_{\phi_1}{}^2 + V_{\phi_2}{}^2 + 2V_{\phi_1} V_{\phi_2} = V_{\phi_3}{}^2; \tag{219}$$

therefore

$$\frac{2}{T}\int_0^T V_{\phi_1} V_{\phi_2} dt = \frac{1}{T}\int_0^T V_{\phi_3}{}^2 dt - \frac{1}{T}\int_0^T V_{\phi_1}{}^2 dt - \frac{1}{T}\int_0^T V_{\phi_2}{}^2 dt.$$

$$= -\bar{V}_\phi{}^2. \tag{220}$$

Substituting in (217), we obtain

$$\bar{V}_\delta{}^2 = 3\bar{V}_\phi{}^2$$

$$\bar{V}_\delta = V_\phi (3)^{\frac{1}{2}}. \tag{221}$$

Performance

Since the transistors are used in the switching mode, the overall efficiency is high (of the order of 80 to 90%). For d.c.-d.c. conversion the increased duty cycle enables the transistors to handle more power than in the push-pull circuit, but there is a corresponding increase in the peak voltage applied to the transistors.

The harmonic content of the output waveform is high, so that attempts to provide a sine-wave output by filtering out harmonics are likely to cause a loss of efficiency of over 30%. Even so, the overall efficiency should still be acceptable for some applications.

REFERENCES

1. Mullard Limited, 1960, *Reference Manual of Transistor Circuits.*
2. M. Dayal, 1964, 'Power rectification with silicon diodes', *Mullard Tech. Commun.*, Vol. 7, No. 68, pp. 230–62.
3. R. N. Hall, 1952, 'Power rectifiers and transistors', *Proc. I.R.E.*, Vol. 40, No. 11, pp. 1512–18.
4. Motorola Semiconductor Products, Incorporated (U.S.A.), 1966, *Silicon Rectifier Handbook.*
5. G. L. Pearson, 1952, 'Silicon p–n junction alloy diodes', *Proc. I.R.E.*, Vol. 40, No. 11, pp. 1348–51.
6. Motorola Semiconductor Products Incorporated (U.S.A.), 1961, *Silicon Zener Diode and Rectifier Handbook.*
7. E. J. Diebold, 1961, 'Semiconductor rectifier characteristics', *Electro-Technol.*, Vol. 67, No. 6, pp. 72–9.
8. E. J. Diebold and D. W. Borst, 1963, 'Semiconductor rectifier diodes and controlled rectifiers', *Electro-Technol.*, Vol. 71, No. 2, pp. 106–12.
9. Mullard Limited, *Mullard Technical Handbook.*
10. General Electric Company (U.S.A.), 1962, *Transistor Manual*, 6th edn.
11. J. H. Tuley, 1960, 'Design of cooling fins for silicon power rectifiers', *Mullard Tech. Commun.*, Vol. 5, No. 44, pp. 118–30.
12. M. Dayal, 1964, 'Rectifier diode operation at kilocycle frequencies', *Mullard Tech. Commun.*, Vol. 8, No. 73, pp. 66–77.
13. D. F. Grollet, 1961, 'Measurement of semiconductor diode transient characteristics', *Mullard Tech. Commun.*, Vol. 5, No. 50, pp. 385–91.
14. J. R. Nowicki, A. Hale, and C. A. Richardson, 1967, 'Voltage regulator (Zener) diodes', *Mullar Tech. Publ.*, No. TP 652.
15. C. Zener, 1934, 'A theory of the electrical breakdown of solid dielectrics', *Proc. Roy. Soc. (London), Ser. A*, Vol. 145, p. 523.
16. A. G. McKay and K. B. McAffe, 1953, 'Electron multiplication in silicon and germanium', *Phys. Rev.*, Vol. 91, p. 1079.
17. F. W. Gutzwiller, 'An introduction to the controlled avalanche silicon rectifier', *Gen. Elect. Co. (U.S.A.), Appl. Note*, No. 200.27, 5/64.
18. F. W. Gutzwiller, 1962, 'Controlled avalanche: a new approach to protecting silicon rectifier diodes against voltage transients', *Direct Current*, pp. 335–8.
19. D. R. Coleman, 1963, 'Silicon avalanche rectifier diodes', *Direct Current*, pp. 268–72.
19. D. R. Coleman and G. Duddridge, 1963, 'Silicon avalanche rectifier diodes', *Direct Current*, pp. 268–72.
20. H. A. Schafft and J. C. French, 1962, 'Second breakdown in transistors', *I.R.E., Trans. Electron. Devices*, Vol. ED-9, pp. 129–36.
21. R. M. Scarlett and W. Schockley, 1963, 'Second breakdown and hot spots in power transistors', *Int. Conv. Record I.E.E.E.*, part 3.
22. J. J. Ebers and J. L. Moll, 1954, 'Large signal behaviour of junction transistors', *Proc. I.R.E.*, Vol. 42 pp. 1761–72.

23. R. Beaufoy and J. J. Sparkes, 1957, 'The junction transistor as a charge controlled device', *A.T.E. Journal*, Vol. 13, No. 4, pp. 310–27. (See also 1957, *Proc. I.R.E.*, Vol. 45, No. 12, pp. 1740–2.)

24. P. James and A. F. Newell, 1960, 'Switching times for alloy junction transistors', *Mullard Tech. Commun.*, Vol. 5 No. 44, pp. 159–71.

25. J. L. Moll, 1954, 'Large-signal transient response of junction transistors', *Proc. I.R.E.*, Vol. 42, pp. 1773–84.

26. R. L. Bright, 1955, 'Junction transistors as switches', *Trans. A.I.E.E. (Commun. and Electron.)*, pp. 111–21.

27. D. F. Dunster, 1965, 'Charge control of switching transistors', *Electron. Equipment News*, pp. 114–20.

28. P. P. Balthasar, 1966, 'Avoid power transistor failure', *Electron. Design*, Vol. 1, pp. 52–6; Vol. 2, pp. 192–7. (See also *BENDIX Corp., Semicond. Div, Appl. Note.*)

29. P. Balthasar, 1965, 'Selecting switching circuit transistors', *Electron. Design News*, (Copyright 1968 by Cahners Publishing Co.) Nov., pp. 72–6 (See also *BENDIX Corp., Semicond. Div., Appl. Note.*)

30. R. M. Mann, 1967, '10 transistor leakage currents', *Electron. Design*, Vol. 15, July, pp. 76 82.

31. H. S. Smith, 1967, 'Switch high loads with power transistors', *Electron. Design*, Vol. 17, Aug. pp. 224–33.

32. Westinghouse Electric Corporation (U.S.A.), 1964, *Silicon Controlled Rectifier Designers' Handbook*.

33. General Electric Company (U.S.A.), 1967, *SRC Manual* 4th edn.

34. J. V. Yonushka, 'Device developments and applications of R.C.A. thyristors', *R.C.A., Appl. Note*, No. ST-3262B, 4/67.

35. T. C. McNulty, 'Device developments and applications of R.C.A. thyristors', *R.C.A., Appl. Note*, No. ST 3963, 4/68.

36. I. Somos, 1961, 'Switching characteristics of silicon power controlled rectifiers, 1. — Turn-on action', *Trans. A.I.E.E. (Commun. and Electron.)*, Part 1, Vol. 80, pp. 320–6.

37. D. W. Borst, 1966, 'Turn-on action in large controlled rectifiers', *Int. Rectifier News*.

38. International Rectifier Company Limited (Great Britain), 1967, 'Silicon thyristors', *I.R. Engng. Inform. Rep.*, No. 204.

39. C. M. Sinclair, 1962, 'Semiconductor survey', *Instrum. Pract.*, Part 1, pp. 1111–15; Part 2, pp. 1368–76; Part 3, pp. 1466–72.

40. D. R. Muss and C. Goldberg, 1963, 'Switching mechanism in the n–p–n–p silicon controlled rectifier', *I.E.E.E. Trans., Electron Devices*, Vol. ED-10, No. 3, pp. 113–20.

41. J. J. Ebers, 1952, 'Four terminal p–n–p–n transistors', *Proc. I.R.E.*, Vol. 40, pp. 1361–6.

42. I. M. Mackintosh, 1958, 'The electrical characteristics of silicon p–n–p–n triodes', *Proc. I.R.E.*, Vol. 46, pp. 1229–35.

43. A. R. Mulica, 1964, 'How to use silicon controlled rectifiers in series or parallel', *Control Engng.*, Vol. 11, No. 5, pp. 95–9.

44. L. Easaki, 1958, 'New phenomenon in narrow germanium p–n junctions', *Phys. Rev.*, Vol. 109, pp. 603–4.

45. H. S. Sommers, Jr., 1959, 'Tunnel diodes as high-frequency devices', *Proc. I.R.E.*, Vol. 47, pp. 120–6.

46. N. Holonyak, Jr., and I. A. Lesk, 1960, 'Gallium arsenide tunnel diodes', *Proc. I.R.E.*, Vol. 48, pp. 1405–9.

47. F. M. Carlson, 'Measurement of tunnel diode parameters', *R.C.A., Appl. Note*, No. SMA. 6.

48. R. C. Sims, E. R. Beck, Jr., and V. C. Kamm, 1961, 'A survey of tunnel-diode digital techniques', *Proc. I.R.E.*, Vol. 49, pp. 136–46.

49. H. R. Kaupp and D. R. Crosby, 1961, 'Calculated waveforms for tunnel diode locked pair', *Proc. I.R.E.*, Vol, 49, pp. 146–54.

50. L. Lesk *et al.*, 1959, 'Germanium and silicon tunnel diodes design, operation ang applications', *I.R.E. WESCON Conv. Record*, Part 3, pp. 9–31.

51. S. Wang, 1961, 'Converter efficiency and power output of a tunnel diode relaxation oscillator', *Proc. I.R.E.*, Vol. 49, pp. 1219–20.

52. H. F. Storm and D. P. Shattuck, 1961, 'Tunnel diode d.c. power converter' presented at *A.I.E.E. Winter Gen. Meet., New York.*

53. A. C. Scott, 1962, 'Symmetrical d.c. converter using 6-a tunnel diodes', *Proc. I.R.E.*, Vol. 50, p. 1851.

54. M. Schuller and W. W. Gartner, 1961, 'Large-signal circuit theory for negative-resistance diodes, in particular tunnel diodes', *Proc. I.R.E.*, Vol. 49, pp. 1268–78.

55. A. Ferendeci, 1962, 'A two-term analytical approsimation of tunnel diode static characteristics', *Proc. I.R.E.*, Vol. 50, pp. 1852–3.

56. K. Turnay, 1962, 'Approximation of the tunnel diode characteristics', *Proc. I.R.E.*, Vol. 50, pp. 202–3.

57. K. Turnay, 1962, 'The maximum power output of the tunnel diode oscillator', *Proc. I.R.E.*, Vol. 50, pp. 2120–1.

58. E. Adler and B. Selikson, 1962, 'A review of the tunnel diode', *Electron. Engng.*, Vol. 34, pp. 8–13, pp. 82–6.

59. J. L. Moll, M. Tanenbaun, J. M. Goldey, and N. Holonyak, 'P–n–p–n transistor switches', *Proc. I.R.E.*, Vol. 44, pp. 1172–82.

60. G. E. McDuffie, Jr., and W. L. Chadwell, 1960, 'An investigation of the dynamic switching properties of four-layer diodes', *Trans. A.I.E.E. (Commun. and Electron.)*, pp. 50–3.

61. J. A. Hoerni and R. N. Noyce, 1958, 'P–n–p–n switches', 1958 *I.R.E. WESCON Conv. Record*, Vol. 2, Part 3, pp. 172–5.

62. A. W. Carlson and R. H. McMahon, 1960, 'P–n–p–n four-layer diodes in switching functions', *Elect. Manuf.*, Jan., pp. 71–8.

63. J. Bliss and D. Zinder, 'Four-layer and curren-limiter diodes reduce circuit cost complexity', *Motorola Semicond. Prod. Inc., Appl. Note*, No. AN-221.

64. Motorola Semiconductor Products Incorporated (U.S.A.), 19--, 'Theory and characteristics of the unijunction transistor', *Motorola Semicond. Prod. Inc., Appl. Note*, No. AN-293.

65. R. H. Crawford and R. T. Dean, 1963, 'Stabilising the unijunction transistor', *Electro-Technol.*, July, pp. 110–12.

66. B. Crawford and R. T. Dean, 1964, 'The unijunction transistor in relaxation circuits'. *Electro-Technol.*, Part 1, Feb., pp. 34–8; Part 2, Mar. pp. 40–5.

67. D. V. Jones, 1966, 'Use the unijunction transistor', *Electron. Design*, Vol. 14, No.12. pp. 114–17.

68. T. P. Sylvan, 'Notes on the application of the silicon unijunction transistor', *Gen. Elect. Co. (U.S.A.), Appl. Note*, Nos. 90.10, 5/61 and 90.10, 5/65.

69. D. V. Jones Unijunction temperature compensation', *Gen. Elect. Co. (U.S.A.), Appl. Note*, No. 90.12, 4/62.

70. General Electric Company (U.S.A.): 1964, 'Unijunction transistors circuits', *Transistor Manual*, 7th edn., Chap. 13.

71. General Electric Company (U.S.A.), 1964, 'Silicon controlled switches', *Transistor Manual*, 7th edn., Chap. 16.

72. H. R. Jones, 1968, 'Miniwatt silicon controlled switch BRY39', *Miniwatt Dig.*, Mar. Apr., pp. 37–41.

73. J. Rozenboom, 1968, 'Diac triggering of thyristors and triacs', *Electron. Appl.*, Vol. 28, No. 3 pp. 85–94.

74. J. M. Neilson, 'Light dimmers using triacs', *R.C.A., Appl. Note*, No. AN-3778.

313

75. J. H. Galloway, 'Using the triac for control of a.c. power', *Gen. Elect. Co., Appl. Note,* No. 200.35, 3/66.
76. F. E. Centry, R.I. Scace, and J. K. Flowers, 1965, 'Bidirectional triode p–n–p–n switches', *Proc. I.E.E.E.,* Vol. 53, No. 4, pp. 355–69.
77. J. V. Yonushka, 'Triac power-control applications', *R.C.A., Appl. Note,* No. AN-3697.
78. P. Coleby, 1963, 'Thermal resistance of semiconductor devices under steady-state conditions', *Mullard Tech. Commun.,* Vol. 7, No. 65, pp. 127–40.
79. Institute of Radio Engineers, 1960, 'I.R.E. standards on solid-state devices: definition of semiconductor terms', *Proc. I.R.E.,* Vol. 48, No. 10, pp. 1772–5.
80. British Standards Institution, 1961, 'Letter symbols for light-current semiconductor devices', *Br. Stand.,* No. 3363 : 1961.
81. O. J. Edwards, 1958, 'Heat sink design', *Mullard Tech. Commun.,* Vol. 3, No. 29, pp. 258–64.
82. H. Sutcliffe, 1961, 'Fluidized beds as constant temperature enclosures', *Electron. Engng.,* Vol. 33, No. 396, pp. 94–5.
83. T. J. Nelson and J. E. Iwersen, 1958, 'Measurement of internal temperature rise of transistors', *Proc. I.R.E.,* Vol. 46, No. 6, pp. 1207–8.
84. H. C. Lin and R. E. Crosby, 1957, 'A determination of thermal resistance of silicon junction devices', *I.R.E. Nat Conv. Record,* Vol. 5, Part 3, pp. 22–5.
85. J. A. van Berge Henegouwen and L. J. Hofland, 1965, 'Cooling systems', *Philips Appl. Inform.,* No. 431.
86. S. Levine, 1965, 'Silicon transistors in power supply design', *Silicon Transistor Corp. Tech. Serv. To Customers,* Vol. 2, No. 1.
87. J. C. Hey, 'A variety of mounting techniques for press fit SCRs and rectifiers', *Gen. Elect. Co., Appl. Note,* No. 200.32, 7/63.
88. R. Greenburg, 'Selecting commercial power transistor heat sinks', *Motorola Semicond. Prod. Inc., Appl. Note,* No. AN-135.
89. A. D. Marquis, 1967, 'How 'hot' are you on thermal rating?', *Electron. Design,* Vol. 23, Nov., pp. 74–5.
90. R. A. Lockett, H. A. Bell, and R. Priston, 1965, 'Thermal resistance of low-power semi-conductor devices under pulse conditions', *Mullard Tech. Commun.,* Vol. 8, No. 76, pp. 146–61.
91. K. E. Mortenson, 1957, 'Transistor junction temperature as a function of time', *Proc. I.R.E.,* Vol. 45, No. 4, pp. 504–13.
92. F. W. Gutzwiller and T. P. Sylvan, 1961, 'Power semiconductor ratings under transient and intermittent loads', *A.I.E.E. Trans. (Commun. and Electron.),* Vol. 19, Part I, pp. 696–706.
93. O. H. Schade, 1943, 'Analysis of rectifier operation', *Proc. I.R.E.,* Vol. 31, No. 7, pp. 341–61.
94. D. L. Waidelich, 1941, 'Diode rectifying circuits with capacitance filters', *Trans. A.I.E.E.,* Vol. 60, pp. 1161–7.
95. B. J. Roman and J. M. Neilson, 'Application of R.C.A. silicon rectifiers to capacitive loads', *R.C.A., Appl. Note,* No. SMA-15.
96. D. B. Corbyn, 1952, 'Special rectifier circuits', *Electron Engng.,* Vol. 24, No. 295, pp. 418–9.
97. E. E. Von Zastrow, 'Capacitor input filter design with silicon rectifier diodes', *Gen. Elect. Co., Appl. Note,* No. 200.30, 8/67.
98. E. E. Von Zastrow, 'The series connection of rectifier diodes', *Gen. Elect. Co., Appl. Note,* No. 200.39, 10/64.
99. N. H. Roberts, 1936, 'The diode as half-wave, full-wave, and voltage-doubling rectifier, with special reference to the voltage output and current input', *Wireless Engr.,* Vol. 13, No. 154, pp. 351–62: No. 155, pp. 423–30.
100. G. J. Tobisch, 1964, 'Parallel operation of silicon diode rectifiers', *Mullard Tech. Commun.,* Vol. 8, No. 73, pp. 78–88.

101. F. W. Gutzwiller, 1958, 'The current-limiting fuse as fault protection for semiconductor rectifiers', *Trans. A.I.E.E.* (*Commun and Electron.*), No. 35, pp. 751–5.

102. C. Le Can, 1959/1960, 'Transient behaviour and fundamental transistor parameters', *Electron. Appl.*, Vol. 20, No. 2, pp. 56–83.

103. G. E. Groves, 1962, 'Surge suppression circuits for semiconductor power rectifiers', *Direct Current*, Mar., pp. 76–8.

104. Tseng-Wu Liao and T. H. Lee, 1966, 'Surge suppression for the protection of solid-state devices', *I.E.E.E. Trans.*, Vol. IGA-2 No. 1, pp. 44–52.

105. F. W. Gutzwiller, 'Overcurrent protection of semiconductor rectifiers', *Gen. Elect. Co.*, *Appl. Note*, No. 200.10, 11/61.

106. Mullard Limited, 1966, 'Idealised rectifier circuit performances', *Mullard Tech. Commun.*, Vol. 9, No. 82, pp. 46–7.

107. M. G. Say, 1963, *The Performance and Design of Alternating Current Machines*, reprint, Pitman, London.

108. C. M. Chen, 'Predicting reverse recovery time of high speed semiconductor junction diodes', *Gen. Elect. Co.*, *Appl. Note*, No. 90.36, 4/62.

109. J. H. Galloway, 'Application of fast recovery rectifiers', *Gen. Elect. Co.*, *Appl. Note*, No. 200.38, 6/65.

110. D. W. Borst and D. Cooper, 1968, 'Application and characterization of a 250 ampere fast recovery rectifier', *Int. Rectifier, Rectifier News*, Winter pp. 2–7.

111. R. Greenburg, 1964, 'Synchronous rectifier using transistor', *Electron. Prod.*, p. 35.

112. L. P. Hunter, 1956, *Handbook of Semiconductor Electronics*, 1st edn., McGraw-Hill, New York, pp. 16–28.

113. L. H. Light and P. M. Hooker, 1956, 'The design and operation of transistor d.c. converters', *Mullard Tech. Commun.*, Vol. 2, No. 17, pp. 157–204.

114. H. H. van Abbe and J. J. Rongen, 1955, 'The design of transistor d.c. converters', *Electron. Appl.*, Vol. 16, No. 2, pp. 59–79.

115. D. L. Johnston, 1954, 'Transistor h.t. generator', *Wireless World*, Vol. 60, No. 10, p. 518.

116. L. H. Light, 1955, 'Transistor power supplies', *Wireless World*, Vol. 61, No. 12, pp. 582–6.

117. W. Hirschmann, 1959, 'Transistor-Eintakt-Zerhacker', *Radio Mentor*, No. 7, Part 1, pp. 526–8; No. 8, Part 2, pp. 613–6; No. 9, Part 3, pp. 713–5.

118. T. Konopinski, 1962, 'The influence of transformer losses on the operation of ringing choke transistor converters', *Direct Current*. Feb., pp. 55–8.

119. T. Konopinski, 1959, 'Charakterystyka obciazenia przetwornic tranzystorowych' (Load characteristics of transistor converter), *Prace Inst. Tele-I Radiotech.*, Vol. 3, pp. 87–96.

120. L. R. Bright, G. F. Pittman, Jr., and G. H. Royer, 1954, 'Transistors as on–off switches in saturable core circuits', *Elect. Manuf.*, Dec., pp. 79–82.

121. G. Grimsdell, 1955, 'The economics of the transistor d.c. transformer', *Electron Engng.*, Vol. 27, No. 328, pp. 268–9.

122. G. H. Royer, 1955, 'A switching transistor d.c.-to-a.c. converter having an output frequency proportional to the d.c. input voltage', *Trans A.I.E.E.* (*Commun. and Electron.*), pp. 322–6.

123. R. L. van Allen, 1955, 'A variable-frequency magnetic-coupled multi-vibrator', *Trans. A.I.E.E.* (*Commun. and Electron.*), pp. 356–61.

124. G. C. Uchrin, 1956, 'Transistor power converter capable of 250 watts d.c. output', *Proc. I.R.E.*, Vol. 44, No. 2, pp. 261–2.

125. K. Chen and A. J. Schiewe, 1956, 'A single-transistor magnetic-coupled oscillator'. *Ttans. A.I.E.E.* (*Commun. and Electron.*), pp. 396–400.

126. W. L. Stephenson, L. P. Morgan, and T. H. Brown, 1959, 'The design of transistor push–pull d.c. convertors', *Electron. Engng.*, Vol. 31, No. 380, pp. 585–9.

127. J. L. Jensen, 1957, 'An improved square-wave oscillator circuit', *I.R.E. Trans.*, Circuit Theory, Vol. CT-4, pp. 276–9.

128. J. R. Nowicki, 1960, 'New high-power d.c. converter circuits', *Mullard Tech. Commun.*, Vol. 5, No. 43, pp. 104–14.

129. J. R. Nowicki, 1961, 'Improved high-power d.c. converters', *Electron. Engng.*, Vol. 33, No. 404, pp. 637–41.

130. H. T. Breece, 'High-speed invertors using silicon power transistors', *R.C.A.*, *Appl. Note*, No. SMA-37.

131. D. T. De Fino, 'A 100-W 18 kHz inverter using RCA-2N5202 silicon power transistors', *R.C.A.*, *Appl. Note*, No. AN-3565.

132. J. Takesuyo, '100 W d.c.-to-d.c. converter operates at 100 kC', *Motorola Semicond. Prod.*, *Inc.*, *Appl. Note*, No. AN-153.

133. Texas Instruments Incorporated (U.S.A.), 1961, '60-cycle germanium power inverters', *Texas Instrum.*, *Inc.*, *Appl. Note*, No. SC-2083.

134. Radio Corporation of America (U.S.A.), 1967, *Silicon Power Circuits Manual*, Tech. Ser., No. SP-50.

135. J. R. Nowicki, 1964, 'D.c. inverters using ADZ11 and ADZ12 transistors', *Mullard Tech. Commun.*, Vol. 7, No. 70, pp. 305–9.

136. K. Wetzel, 1959, 'Frequenzkonstanter Leistungszerhacker', *Radio Mentor*, No. 12, pp. 958–9.

137. A. Haug, 1959, 'Transistor-Gegentakt-Zerhackern', *Radio Mentor*, No. 12, pp. 964–6.

138. J. R. Nowicki, *Br. Pat. Appl.*, No. 21052/68.

139. J. R. Nowicki, 1961, 'D.c. inverter for fluorescent lamp', *Mullard Tech. Commun.*, Vol. 5, No. 47, pp. 276–85.

140. J. R. Nowicki, 1962, 'A d.c. invertor with *CR* timing', *Electron. Engng.*, Vol. 34, No. 413, pp. 464–8.

141. T. Henenkamp and J. J. Wilting, 1958/1959, 'Transistor d.c. converters for fluorescent-lamp power supplies', *Philips Tech. Rev.*, Vol. 20, No. 12, pp. 362–6.

142. W. Elenbaas, 1959, *Fluorescent Lamps and Lighting*, Philips Tech. Library, p. 115.

143. J. R. Nowicki, 1968, 'Smaller d.c. converters and inverters', *Wireless World*, Vol. 74, No. 389, pp. 38-42. (See also 1968, *Wireless World*, Vol. 74, No. 390, p. 83.)

144. W. L. Stephenson, 1958, 'A four-transistor d.c. converter circuit for use with relatively high-voltage supplies', *Mullard Tech. Commun.*, Vol. 4, No. 36, pp. 191–2.

145. J. R. Nowicki, *Br. Pat. Appl.*, No. 18158/68.

146. J. R. Nowicki, 1963, 'D.c. inverter for electric shavers', *Wireless World*, Vol. 69, No. 12, pp. 624–7. (See also 1964, *Wireless World*, Vol. 70, No. 1, p. 23.)

147. J. M. Marzolf, 1961, 'Tunnel diode static inverter', *Naval Res. Lab.*, *Rep.*, No. 5706.

148. F. M. Carlson and P. D. Gardner, 1963, 'Tunnel diode converters', *Proc. Power Sources Conf.*, pp. 158–62.

149. R. Feryszka and P. Gardner, 1965, 'Tunnel diode low-input-voltage inverter', *Proc. Power Sources Conf.*, pp. 116–20. (See also *R.C.A.*, *Publ.*, No. ST-2914.)

150. R. Feryszka and P. Gardner, 1965, 'Push-pull saturated core tunnel diode inverters', *R.C.A. Rev.*, Vol. 26.

151. A. G. Milnes, 1955, 'Phase locking of switching-transistor converters for polyphase power supplies'. *A.I.E.E. Trans. (Commun. and Electron.)*, No. 21, pp. 587–92.

152. W. E. Jewet and P. L. Schmidt, 1959, 'A more stable three-phase transistor core power inverter', *A.I.E.E. Trans. (Commun. and Electron.)*, Vol. 78, pp. 686–91.

153. C. H. R. Campling and J. A. Bennet, 1961, 'Self-locking polyphase transistor magnetic inverters', *A.I.E.E. Trans. (Commun. and Electron.)*, pp. 26–33.

154. Motorola Semiconductor Products Incorporated (U.S.A.), 'The ABC's of d.c.-to-a.c. inverters', *Motorola Semicond. Prod. Inc.*, *Appl. Note*, No. AN-222.

155. W. L. Stephenson, 1960, 'Transistor three-phase d.c.–a.c. inverter', *Mullard Tech. Commun.*, Vol. 5, No. 42, pp. 78–80.

INDEX